This book is to be returned on or before
the last date stamped below.

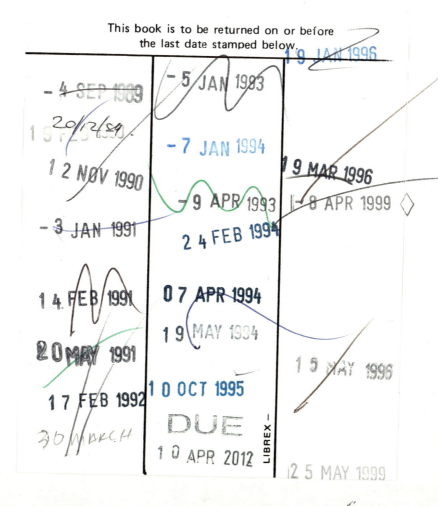

BASIC PROGRAMS FOR CHEMICAL ENGINEERS

BASIC PROGRAMS FOR CHEMICAL ENGINEERS

Dennis Wright

VNR VAN NOSTRAND REINHOLD COMPANY ————————————————————— New York

Disclaimer

Extreme care has been taken in preparing the computer programs listed in this volume. Extensive testing and checking have been performed to insure accuracy and effectiveness of computer solutions. However, neither the author nor the publisher shall be held responsible or liable for any damages resulting in connection with or arising from the use of any of the programs in this book.

Manufactured in the United States of America

Published by Van Nostrand Reinhold Company Inc.
115 Fifth Avenue
New York, New York 10003

Van Nostrand Reinhold Company Limited
Molly Millars Lane
Wokingham, Berkshire RG11 2PY, England

Van Nostrand Reinhold
480 La Trobe Street
Melbourne, Victoria 3000, Australia

Macmillan of Canada
Division of Canada Publishing Corporation
164 Commander Boulevard
Agincourt, Ontario M1S 3C7, Canada

15 14 13 12 11 10 9 8 7 6 5 4 3 2

Library of Congress Cataloging-in-Publication Data

Wright, Dennis.
  Basic programs for chemical engineers.

  Includes index.
    1. Chemical engineering—Computer programs.
2. BASIC (Computer program language)   I. Title.
TP184.W75 1986   660.2′8′00285526   85-22712
ISBN 0-442-29296-1

PREFACE

The microcomputer has put a vast amount of computational power in the hands of the practicing chemical engineer. However, a microcomputer is of little use unless there are programs available to solve chemical engineering problems. In this book, I have put together a collection of BASIC programs that will help the practicing engineer be more productive and able to solve complex problems that are normally handled on mainframe computers.

The plant engineer will find the book particularly useful. The plant engineer is called upon to investigate problems that range from simple trouble shooting to the detailed design of complex chemical plants. The larger projects are usually add-on jobs to the regular duties of keeping a chemical plant running. In today's business climate, answers to problems must be obtained quickly and accurately. The computer is capable of testing hypothesis, thereby allowing engineers to evaluate alternative solutions to problems quickly and provide answers to management's questions that invariably shift like the sands in a desert.

The programs do much more than a programmable calculator such as TI-59 or HP 97/67. Many of the programs contain large amounts of primary data allowing engineers to concentrate on solving problems instead of trying to find obscure data in some handbook and then entering them into numerous calculator memory registers. The programs ask for needed data in common engineering units or in units that are popular in handbooks such as the *CRC Manual* and Lang's *Handbook of Chemistry*. The programs' outputs provide clear documentation of the conditions that the calculated results are based on, which is something that programmable calculators cannot do. The programs self-documenting. Programs that request data from user-entered data statements are clearly illustrated by examples.

The programs are written in a simple version of TRS-80 Model III BASIC. I have used commands that are common to most, if not all, popular versions of Microsoft BASIC. Each program requires less than 16K of memory, and all calculations are done in single precision (7 decimals).

You should not find it difficult to adapt the TRS-80 syntax to your particular version of BASIC. The symbol for exponentiation "[" used by the TRS-80 can be replaced with "^" when the Programs are intended to be run on the IBM PC or the Apple II computers. The Programs employing "PRINT USING" statements are compatible with IBM PC BASIC, but are not compatible with the Apple II because Apple does not have an equivalent statement.

Finally, I urge the user to use your imagination when it comes to applying the programs to particular problems. Don't be afraid to make changes in program logic that will customize the program to a local situation. I hope that you will find the programs as useful as they have been for me.

DENNIS WRIGHT

ACKNOWLEDGMENTS

I deeply wish to thank the people who helped me put this book together. I particularly want to thank my wife Jane, my friend Jane, my secretary Billi, and Connie who typed most of the manuscript.

CONTENTS

Preface / v

1. **MATHEMATICS** / 1
Data Reduction / 1
Roots of Polynomials / 18
Differential Equation / 20
Systems of Non-linear Equations / 26

2. **HEAT TRANSFER** / 32
Shell and Tube Heat Exchangers / 32
Double Pipe Heat Exchangers / 45

3. **THERMODYNAMICS** / 63
Chemical Reaction Stoichiometry / 63
Chemical Equilibrium / 67
Vapor Liquid Equilibrium / 78
Gas Solubility / 88

4. **MASS TRANSFER** / 96
Multicomponent Distillation / 96
Sieve Plate Efficiency / 107
Packed Tower Design / 118
Sieve Plate Hydraulics / 123

5. **COST ESTIMATION AND ECONOMICS** / 136
Flow Sheet Estimater / 136
Preliminary Project Economics / 145
Engineering Economics / 162
TEMA Heat Exchanger Estimates / 169

6. PHYSICAL PROPERTIES OF PURE SUBSTANCES / 181
Critical Temperature / 182
Critical Pressure / 187
Critical Volume / 192
Boiling Point, Vapor Pressure, Heat of Vaporization / 197
Heat Capacity of Gases and Liquids / 203
Viscosity and Thermal Conductivity of Gases / 215
Liquid Viscosity and Density / 225
Liquid Thermal Conductivity and Surface Tension / 235

7. POLLUTION CONTROL / 242
Pollution Dispersion / 242
Cyclone Evaluation and Design / 248
Venturi Scrubbers / 264
Packed Tower Absorbers and Strippers / 273

8. MISCELLANEOUS / 283
Liquid and Gas Flow Meters / 283
Tank Volume Calculator / 292
Data Encryption / 300
Economical Insulation Thickness / 304

APPENDIX / 313
Vapor Pressures of Organic Compounds / 313
Solubility Parameters of Various Liquids at 25°C / 328
Group Contributions to Partial Solubility Parameters / 334
Estimated Tube Counts / 334
Estimated Installed Insulation Costs / 337

INDEX / 339

BASIC PROGRAMS FOR CHEMICAL ENGINEERS

1
MATHEMATICS

DATA REDUCTION

Engineering data reduction can be a time-consuming task. It is essential that the practicing engineer have available quick and easy methods for data reduction. Regression techniques and data plotting are indispensable tools. REGRESS is a program package that allows you to analyze data with reciprocal, linear, exponential, log linear, parabolic, or geometric regressions. Data can be entered through the key board, from data files on tape, or as data statements starting at line 3200. When using data statements, enter the X value first, Y value next. For example,

3200 DATA X1,Y1,X2,Y2, . . .

The program allows you to change regression formula and add, delete, or change data. After you are through manipulating the data, data can be plotted or saved to tape.

EXAMPLE

READY

RUN

1. ENTER DATA FROM KEYBOARD

2. ENTER DATA FROM DATA STATEMENTS

Arkin, H., Colton, R., *Statistical Methods,* 5th Ed, Barnes and Noble, 1970, New York, N.Y.
Poole, L., Borchers, M., *Some Common Basic Programs,* 3rd Ed, Osborne McGraw-Hill, 1979, Berkeley, Cali.

3. ENTER DATA FROM TAPE

4. LIST DATA

5. CHANGE DATA

6. ADD DATA

7. DELETE DATA

8. RUN REGRESSIONS AND PLOT

9. SAVE DATA TO TAPE

ENTER CHOICE(0-9, ENTER 0 TO END):?   **2** [ENTERED DATA ARE
SHOWN IN **BOLD.**]

HOW MANY DATA PAIRS:?   **5**

1. ENTER DATA FROM KEYBOARD

2. ENTER DATA FROM DATA STATEMENTS

3. ENTER DATA FROM TAPE

4. LIST DATA

5. CHANGE DATA

6. ADD DATA

7. DELETE DATA

8. RUN REGRESSIONS AND PLOT

9. SAVE DATA TO TAPE

ENTER CHOICE(0-9, ENTER 0 TO END):?   **4**

LIST DATA TO LINE PRINTER (Y/N):?   **N**
LIST DATA TO SCREEN (Y/N):?   **Y**

NO.	X-VALUE	Y-VALUE
1	.1	.25
2	.4	.55
3	.6	.65
4	.9	.99

NO.	X-VALUE	Y-VALUE
5	.25	.35
6	0	0
7	0	0
8	0	0
9	0	0
10	0	0
11	0	0
12	0	0
13	0	0

1. ENTER DATA FROM KEYBOARD

2. ENTER DATA FROM DATA STATEMENTS

3. ENTER DATA FROM TAPE

4. LIST DATA

5. CHANGE DATA

6. ADD DATA

7. DELETE DATA

8. RUN REGRESSIONS AND PLOT

9. SAVE DATA TO TAPE

ENTER CHOICE(0–9, ENTER 0 TO END):?   8

1. LINEAR REGRESSION

2. EXPONENTIAL REGRESSION

3. GEOMETRIC REGRESSION

4. RECIPROCAL REGRESSION

5. PARABOLIC REGRESSION

6. LOG-LINEAR REGRESSION

7. PLOT DATA

ENTER CHOICE(0-7, ENTER 0 TO RETURN TO MENU)?   1

LINEAR REGRESSION

$Y = .144923 + .917949X$

COEFFICIENT OF DETERMINATION $(R**2) = .986036$

COEFFICIENT OF CORRELATION =   .992994

STANDARD ERROR OF ESTIMATE =   .0393862

INTERPOLATE-(Y/N)?   N

1. LINEAR REGRESSION

2. EXPONENTIAL REGRESSION

3. GEOMETRIC REGRESSION

4. RECIPROCAL REGRESSION

5. PARABOLIC REGRESSION

6. LOG-LINEAR REGRESSION

7. PLOT

ENTER CHOICE(0-7, ENTER 0 TO RETURN TO MENU)?   7

PLOTTING ROUTINE

X-AXIS:LEFT END POINT, RIGHT ENDPOINT, INCREMENT

?   0,1,.05

Y-AXIS:LOWER ENDPOINT, UPPER END POINT, INCREMENT

?   0,1,.025

```
INTERSECTION OF AXES AT ( 0 , 0 )
X-AXIS RANGE= 0 - 1  INCREMENT= .05
Y-AXIS RANGE= 0 - 1  INCREMENT= .025

+++++++++++++++++++++++++++++++++++++++++++++++++Y-AXIS
+
+           *
+
+
+               *
+
+
+                     *
+
+
+
+                         *
+
+
+
+
+                                     *
+
+
X-AXIS
```

1. LINEAR REGRESSION

2. EXPONENTIAL REGRESSION

3. GEOMETRIC REGRESSION

4. RECIPROCAL REGRESSION

5. PARABOLIC REGRESSION

6. LOG-LINEAR REGRESSION

7. PLOT DATA

ENTER CHOICE(0-7, ENTER 0 TO RETURN TO MENU)? 0

1. ENTER DATA FROM KEYBOARD

2. ENTER DATA FROM DATA STATEMENTS

3. ENTER DATA FROM TAPE

4. LIST DATA

5. CHANGE DATA

6. ADD DATA

7. DELETE DATA

8. RUN REGRESSIONS AND PLOT

9. SAVE DATA TO TAPE

ENTER CHOICE(0–9, ENTER 0 TO END):?   0

PROGRAM EXECUTION TERMINATED

READY

```
   LISTING FOR BASIC PROGRAM "REGRESS"

10 CLS:DIM X(50),Y(50),X1(50),Y1(50)

20 PRINT"REGRESSION AND PLOTTING ROUTINE"

30 PRINT"1. ENTER DATA FROM KEYBOARD"

40 PRINT"2. ENTER DATA FROM DATA STATEMENTS"

50 PRINT"3. ENTER DATA FROM TAPE"

60 PRINT"4. LIST DATA"

70 PRINT"5. CHANGE DATA"

80 PRINT"6. ADD DATA"

90 PRINT"7. DELETE DATA"

100 PRINT"8. RUN REGRESSIONS AND PLOT"

110 PRINT"9. SAVE DATA TO TAPE"

120 PRINT

130 INPUT"ENTER CHOICE(0-9, ENTER 0 TO END):";Z

140 IF Z=0 THEN 170

150 ON Z GOSUB 2020 ,2120 ,2180 ,2260 ,2490 ,2630 ,2740 ,2980 ,3110

160 CLS:GOTO 30

170 CLS:PRINT"PROGRAM EXECUTION TERMINATED"
```

```
180 END

190 SX=0:SY=0:X2=0:Y2=0:XY=0

200 FOR I=1 TO N

210 IF C=1 THEN GOTO 280

220 IF C=2 THEN GOSUB 820    :GOTO 290

230 IF C=3 THEN GOSUB 840    :GOTO 290

240 IF C=4 THEN GOSUB 860    :GOTO 290

250 IF C=5 THEN GOSUB 880    :GOTO 290

260 IF C=6 THEN GOSUB 900    :GOTO 290

270 IF C=7 THEN 1050

280 X=X1(I):Y=Y1(I)

290 SX=SX+X

300 SY=SY+Y

310 X2=X2+X[2

320 Y2=Y2+Y[2

330 XY=XY+X*Y

340 NEXT I

350 B1=(N*XY-SY*SX)/(N*X2-SX[2)

360 A1=(SY-B1*SX)/N

370 LPRINT:LPRINT:LPRINT

380 ON C GOSUB 600  ,630  ,660  ,700  ,740  ,780

390 Z1=(N*XY-SX*SY)[2

400 Z2=(N*X2-SX[2)*(N*Y2-SY[2)

410 R2=Z1/Z2

420 LPRINT:LPRINT

430 Z3=(Y2-SY[2/N)*(1-R2)

440 LPRINT"COEFFICENT OF DETERMINATION (R**2)=";R2
```

Adapted from Poole, L., Borchers, M. *Some Common Basic Programs,* 3rd Ed., Osborne McGraw-Hill. With permission of Osborne McGraw-Hill, 1979.

```
450 LPRINT"COEFFICENT OF CORRELATION=          ";SQR(R2)

460 LPRINT"STANDARD ERROR OF ESTIMATE=          ";SQR(Z3/(N-2))

470 CLS

480 INPUT"INTERPOLATE-(Y/N)";A$

490 IF A$="N" THEN 590

500 LPRINT:LPRINT

510 LPRINT"INTERPOLATION"

520 LPRINT"X",,"Y"

530 LPRINT"----------------------------------------------------"

540 PRINT"INTERPOLATION"

550 INPUT"X-VALUE........9999 TO END";X

560 IF X=9999 THEN 590

570 ON C GOSUB 920 ,940 ,960 ,980 ,1000 ,1020

580 GOTO 550

590 RETURN

600 LPRINT"LINEAR REGRESSION"

610 LPRINT"Y=";A1;"+";B1;"X"

620 RETURN

630 LPRINT"EXPONENTIAL REGRESSION"

640 LPRINT"A=";EXP(A1);"    ";"B=";B1

650 RETURN

660 LPRINT"GEOMETRIC REGRESSION"

670 LPRINT"Y=A*X**B"

680 LPRINT"A=";EXP(A1);"    ";"B=";B1

690 RETURN

700 LPRINT"RECIPORCAL REGRESSION"

710 LPRINT"Y=A+B*1/X"

720 LPRINT"A=";A1;"    ";"B=";B1
```

```
 730 RETURN

 740 LPRINT"PARABOLIC REGRESSION"

 750 LPRINT"Y=A+B*X**2"

 760 LPRINT"A=";A1;"        ";"B=";B1

 770 RETURN

 780 LPRINT"LOG-LINEAR REGRESSION"

 790 LPRINT"Y=A+B*LN(X)"

 800 LPRINT"A=";A1;"        ";"B=";B1

 810 RETURN

 820 Y=LOG(Y1(I)):X=X1(I)

 830 RETURN

 840 X=LOG(X1(I)):Y=LOG(Y1(I))

 850 RETURN

 860 X=1/X1(I):Y=Y1(I)

 870 RETURN

 880 X=X1(I)[2:Y=Y1(I)

 890 RETURN

 900 X=LOG(X1(I)):Y=Y1(I)

 910 RETURN

 920 LPRINT X,,A1+B1*X

 930 RETURN

 940 LPRINT X,,EXP(A1)*EXP(X*B1)

 950 RETURN

 960 LPRINT X,,EXP(A1)*X[B1

 970 RETURN

 980 LPRINT X,,A1+B1*1/X

 990 RETURN

1000 LPRINT X,,A1+B1*X[2
```

```
1010 RETURN

1020 LPRINT X,,A1+B1*LOG(X)

1030 RETURN

1040 REM-PLOTTING ROUTINE FROM "SOME COMMON BASIC PROGRAMS-3RD EDTION",
     LON POOLE, MARY BORCHERS, WITH PERMISSION OSBORNE-MC GRAW HILL
     BOOKS @1979

1050 CLS:PRINT"PLOTTING ROUTINE"

1060 PRINT"X-AXIS:LEFT END POINT,RIGHT ENDPOINT,INCREMENT"

1070 INPUT A1,A2,A3

1080 PRINT"Y-AXIS:LOWER ENDPOINT,UPPER END POINT,INCREMENT"

1090 INPUT B1,B2,B3

1095 LPRINT:LPRINT:LPRINT

1100 LPRINT"INTERSECTION OF AXES AT (";A1;",";B1;")"

1110 LPRINT"X-AXIS RANGE=";A1;"-";A2;" INCREMENT=";A3

1120 LPRINT"Y-AXIS RANGE=";B1;"-";B2;" INCREMENT=";B3

1130 B2=(B2-B1)/B3

1140 IF B2<=80 THEN 1170

1150 PRINT"Y RANGE TOO LARGE"

1160 GOTO 1080

1170 FOR I=1 TO N

1180 X(I)=INT((X1(I)-A1)/A3+.5)

1190 Y(I)=INT((Y1(I)-B1)/B3+.5)

1200 NEXT I

1210 Y(N+1)=INT(B2+.5)+1

1220 X(N+1)=INT((A2-A1)/A3+.5)+1

1230 LPRINT:LPRINT

1240 FOR J=1 TO N

1250 FOR I=1 TO N-J

1260 A=X(I)
```

```
1270 B=Y(I)

1280 C=X(I+1)

1290 D=Y(I+1)

1300 IF A<C THEN 1350

1310 X(I)=C

1320 Y(I)=D

1330 X(I+1)=A

1340 Y(I+1)=B

1350 NEXT I

1360 NEXT J

1370 T=1

1380 FOR P=0 TO N-1

1390 IF X(P+1))=0 THEN 1410

1400 NEXT P

1410 FOR I=0 TO INT((A2-A1)/A3+.5)

1420 T=T+P

1430 P=0

1440 IF T)N THEN 1460

1450 IF X(T)=I THEN 1510

1460 IF I=0 THEN 1490

1470 LPRINT"+";

1480 GOTO 1980

1490 S=N+1

1500 GOTO 1860

1510 FOR L=T TO N

1520 IF X(L))X(T) THEN 1550

1530 P=P+1

1540 NEXT L
```

```
1550 IF P=1 THEN 1670

1560 N1=0

1570 N1=N1+1

1580 IF N1)P THEN 1670

1590 FOR L=1 TO P-N1

1600 D=Y(T+L-1)

1610 B=Y(T+L)

1620 IF D(=B THEN 1650

1630 Y(T+L-1)=B

1640 Y(T+L)=D

1650 NEXT L

1660 GOTO 1570

1670 FOR L=0 TO P-1

1680 Z=Y(T+L)

1690 IF Z)=0 THEN 1710

1700 NEXT L

1710 IF I=0 THEN 1850

1720 IF Z=0 THEN 1740

1730 LPRINT"+";

1740 IF L=P-1 THEN 1810

1750 FOR J=L TO P-1

1760 IF Z)B2 THEN 1980

1770 IF Y(T+J)=Z THEN 1800

1780 LPRINT TAB(Z);"*";

1790 Z=Y(T+J)

1800 NEXT J

1810 IF Z(0 THEN 1980

1820 IF Z)B2 THEN 1980
```

```
1830 LPRINT TAB(Z);"*";

1840 GOTO 1980

1850 S=T+L

1860 FOR J=0 TO B2

1870 IF Y(S)<>J THEN 1950

1880 LPRINT "*";

1890 FOR K=S TO T+P-1

1900 IF Y(K)=Y(S) THEN 1930

1910 S=K

1920 GOTO 1960

1930 NEXT K

1940 GOTO 1960

1950 LPRINT"+";

1960 NEXT J

1970 LPRINT"Y-AXIS";

1980 LPRINT

1990 NEXT I

2000 LPRINT"X-AXIS"

2010 RETURN

2020 REM- DATA FROM KEYBOARD

2030 CLS:INPUT"HOW MANY DATA PAIRS:";N

2040 FOR I=1 TO N

2050 PRINT"X(";I;")=";

2060 INPUT X1(I)

2070 PRINT"Y(";I;")=";

2080 INPUT Y1(I)

2090 NEXT I

2100 RETURN
```

```
2110 RETURN

2120 REM-DATA FROM DATA STATEMENTS

2130 CLS:INPUT"HOW MANY DATA PAIRS:";N

2140 FOR I=1 TO N

2150 READ X1(I),Y1(I)

2160 NEXT I

2170 RETURN

2180 REM- DATA FROM TAPE

2190 CLS:PRINT"PRESS PLAY BUTTON"

2200 INPUT A$

2210 INPUT #-1,N

2220 FOR I=1 TO N

2230 INPUT #-1,X1(I),Y1(I)

2240 NEXT I

2250 RETURN

2260 REM-LIST DATA TO SCREEN OR TO LP

2270 INPUT"LIST DATA TO LINE PRINTER(Y/N):";A$

2280 IF A$="Y" THEN 2440

2290 INPUT"LIST DATA TO SCREEN(Y/N):";A$

2300 IF A$="N" THEN RETURN

2310 CLS

2320 J=INT(N/14+1):N1=0

2330 FOR I=1 TO J

2340 PRINT"NO.","X-VALUE","Y-VALUE"

2350 PRINT"---------------------------------------------"

2360 FOR I1=1 TO 13

2370 N1=N1+1

2380 IF X1(I1)=1E38 THEN X1(I1)=0:Y1(I1)=0
```

```
2390 PRINT N1,X1(I1),Y1(I1)

2400 NEXT I1

2410 INPUT A$

2420 NEXT I

2430 RETURN

2440 LPRINT"NO.","X-VALUE","Y-VALUE"

2450 FOR I=1 TO N

2460 LPRINT I,X1(I),Y1(I)

2470 NEXT I

2480 RETURN

2490 REM-CHANGE DATA

2500 CLS:INPUT"LIST DATA(Y/N):";A$

2510 IF A$="Y" THEN GOSUB 2260

2515 CLS

2520 INPUT"HOW MANY CHANGES:";C1

2530 FOR I=1 TO C1

2540 INPUT"CHANGE WHICH DATA PAIR:";C2

2550 PRINT"OLD X=";X1(C2)

2560 PRINT"OLD Y=";Y1(C2)

2570 PRINT"----------------------------"

2580 INPUT"NEW X=";X1(C2)

2590 INPUT"NEW Y=";Y1(C2)

2600 CLS

2610 NEXT I

2620 RETURN

2630 REM-ADD DATA

2640 CLS

2650 INPUT"HOW MANY ADDITIONS:";N2
```

```
2660 FOR I=1 TO N2

2670 PRINT"X(";N+I;")=";

2680 INPUT X1(I+N)

2690 PRINT"Y(";I+N;")=";

2700 INPUT Y1(I+N)

2710 NEXT I

2720 N=N+N2

2730 RETURN

2740 REM-DELETE DATA

2750 CLS

2760 INPUT"LIST DATA(Y/N):";A$

2770 IF A$="Y" THEN GOSUB 2260

2775 CLS

2780 INPUT"HOW MANY DELETIONS:";N3

2790 N4=0

2800 FOR I=1 TO N3

2810 INPUT"DELETE WHICH DATA PAIR:";C3

2820 PRINT"X(";C3;")=";X1(C3)

2830 PRINT"Y(";C3;")=";Y1(C3)

2840 INPUT"DELETE(Y/N):";A$

2850 IF A$="Y" THEN X1(C3)=1E38:N4=N4+1

2860 NEXT I

2870 REM-SORT DATA IN ASCENDING ORDER

2880 FOR I=1 TO N

2890 FOR J=1 TO N-1

2900 IF X1(J))=X1(J+1) THEN 2920

2910 GOTO 2950

2920 T1=X1(J+1):T2=Y1(J+1)
```

```
2930 X1(J+1)=X1(J):Y1(J+1)=Y1(J)

2940 X1(J)=T1:Y1(J)=T2

2950 NEXT J

2960 NEXTI:N=N-N4

2970 RETURN

2980 REM-RUN REGRESSIONS AND PLOT

2990 CLS

3000 PRINT"1. LINEAR REGRESSION"

3010 PRINT"2. EXPONENTIAL REGRESSION"

3020 PRINT"3. GEOMETRIC REGRESSION"

3030 PRINT"4. RECIPORCAL REGRESSION"

3040 PRINT"5. PARABOLIC REGRESSION"

3050 PRINT"6. LOG-LINEAR REGRESSION"

3060 PRINT"7. PLOT"

3070 INPUT"ENTER CHOICE(0-7, ENTER 0 TO RETURN TO MENU)";C

3080 IF C=0 THEN RETURN

3090 GOSUB 190

3100 GOTO 2980

3110 REM-SAVE DATA TO TAPE

3120 CLS

3130 PRINT"PRESS RECORD BUTTON"

3140 INPUT A$

3150 PRINT #-1,N

3160 FOR I=1 TO N

3170 PRINT #-1,X1(I),Y1(I)

3180 NEXT I

3190 RETURN

3200 DATA .10,.25,.40,.55,.60,.65,.90,.99,.25,.35
```

ROOTS OF NONLINEAR FUNCTIONS

POLY is a program that uses first derivatives to obtain the roots to a non-linear function. You enter the function starting at line 300. The program prompts you for an initial guess, a damping factor, the maximum number of iterations, the relative error, and the step size used to calculate the first derivative. A damping factor is used to keep the computer-generated approximations from oscillating wildly past a true root. The variable used by the program is the variable X.

The example used to illustrate the program is the solution to the polynomial equation shown below:

$$F = 1 + 3X^2 - 5X^3$$

The solution to the equation computed to six places is $X = .866425$. The function value at this point is $F = -1.95503E - 5$.

Equations for POLY

$$X_{n+1} = X_n - DF*F/DERV$$
$$DERV = (F1 - F2)/ST$$
$$E = ABS ((X_{n+1} - X_n)/Xn)$$

Nomenclature for POLY

VARIABLE	DESCRIPTION
F, F1, F2	Function value at point X, X + ST
DERV	First Derviative at point X_n
DF	Damping Factor (0 − 1)
X_n	Value of Approximate Root
X_{n+1}	Improved Estimate
ST	Step Size
E	Error Tolerance

EXAMPLE

READY

LOAD "POLY'

Thomas, G., *Calculus and Analytic Geometry,* 4th Ed., Addison-Wesley, 1969, Reading, Mass., pp. 324–327.

READY

RUN

ROOTS OF NONLINEAR EQUATIONS

INITIAL GUESS=? **1**

DAMPING FACTOR(0-1.0)=? **.25**

MAX NO. OF ITERATIONS=? **100**

RELATIVE ERROR=? **1E-6**

STEP SIZE=? **.0001**

*****PROGRAM RUNNING*****

SOLUTION TO SYSTEM

X=.866425   F=−1.95503E-05

READY

```
    LISTING FOR BASIC PROGRAM "POLY"

 10 CLS
 20 PRINT"ROOTS OF NONLINEAR EQUATIONS"
 30 INPUT"INITIAL GUESS=";X
 40 INPUT"DAMPING FACTOR(0-1.0)=";DF
 50 INPUT"MAX NO. OF ITERATIONS=";MAX
 60 INPUT"RELATIVE ERROR=";E
 70 INPUT"STEP SIZE=";ST
 80 CLS
 90 PRINT"*****PROGRAM RUNNING*****"
100 TEMP=X:N=0
110 X=X+ST
120 GOSUB 290
130 F1=F
140 X=X-ST
```

```
150 GOSUB 290

160 F2=F

170 DERV=(F1-F2)/ST

180 X=X-DF*F/DERV

190 IF ABS((TEMP-X)/X))<=E THEN 250

200 N=N+1

210 IF N=>MAX THEN 230

220 TEMP=X: GOTO 110

230 CLS

240 PRINT"NO CONVERGENCE":GOTO 270

250 CLS

260 PRINT"SOLUTION TO SYSTEM"

270 PRINT"X=";X,"F=";F

280 END

290 REM-ADD FUNCTION HERE......USE FORM F=X[2+X[3...ETC.

300 F=1+3*X[2-5*X[3

310 RETURN
```

SIMULTANEOUS DIFFERENTIAL EQUATIONS

Integral calculus is used everyday by the practicing chemical engineer. DIF-FEQ uses a fourth-order Runge-Kutta routine to solve one or several ordinary differential equations. The program is especially useful for mathematically modeling dynamic systems. The program is best illustrated by an example.

PROBLEM: The following reactions are being carried out in a plug flow reactor:

$$A + B === >C \qquad K1 = 1.00$$
$$C + B === >D \qquad K2 = .100$$

Predict the product distribution as a function of residence time.

SOLUTION: The differential equations to be solved are

1. F1 = − K1(A)(B)
2. F2 = − K1(A)(B) − K2 (C)(B)
3. F3 = − F1 − K2(B)(C)
4. F4 = K2(B)(C)

We will now assign variables to A, B, C, and D:

X(1) = A,   X(2) = B,   X(3) = C,   X(4) = D

The functions are now added, starting at line 5000, using the following format:

5000 K1 = 1.00:K2 = .100

5010 F(1) = −K1∗X(1)∗X(2)

5020 F(2) = F(1) − K2∗X(2)∗X(3)

5030 F(3) = −F(1) − K2∗X(2)∗X(3)

5040 F(4) = K2∗X(2)∗X(3)

5050 RETURN

After typing RUN, the program prompts you for

1. Number of equations
2. Step size
3. Print interval
4. Time interval
5. Initial values for each equation

The computer's output is given in the following example. The computer results show that the conversion of compound A is about 89.82%. Compound B is virutally consumed after 2 hr. The conversion of compound C to compound D is only 11.5%.

Equations for DIFFEQ

$$X1(C)_{n+1} = X1(C)_n + [R(1,C) + 2R(2,C) + 2R(3,C) + R(4,C)]/6$$
$$R(1,C) = DT∗F(C) \text{ at } X1(C)_n$$

Milne, W., *Numerical Solution of Differential Equations*, Dover Publications, 1970, New York, N.Y., pp. 72–73.

$$R(2,C) = DT*F(C) \text{ at } X1(C)_n + .5R(1,C)$$
$$R(3,C) = DT*F(C) \text{ at } X1(C)_n + .5R(2,C)$$
$$R(4,C) = DT*F(C) \text{ at } X1(C)_n + .5R(3,C)$$

Nomenclature for DIFFEQ

VARIABLE	DESCRIPTION
$X1(C)_n$	Integrated value at T
$X1(C)_{n+1}$	Integrated value at T + DT
$F(C)$	Differential equation
DT	Step size
$R(N,C)$	Runge-Kutta derivations

EXAMPLE

READY

RUN

SIMULTANEOUS LINEAR DIFF EQUATIONS

NO. OF EQUATIONS = ?   4

TIME INTERVAL, STEP SIZE = ?   3.,.10

INITIAL VALUE OF X(1) = ?   10

INITIAL VALUE OF X(2) = ?   10

INITIAL VALUE OF X(3) = ?   0

INITIAL VALUE OF X(4) = ?   0

PRINT INTERVAL = ?   10

****PROGRAM RUNNING****

CALCULATIONS COMPLETED.....PRINTING ANSWERS

T = 1

X(1) = 1.22508

X(2) = .394006

X(3) = 7.59495

X(4) = 1.03249

T = 2

X(1) = 1.01832

X(2) = .06181

X(3) = 7.65776

X(4) = 1.16721

T = 3

X(1) = .987994

X(2) = .0105805

X(3) = 7.66462

X(4) = 1.18939

PROGRAM EXECUTION TERMINATED

READY

```
    LISTING FOR BASIC PROGRAM "DIFFEQ"

10 DIM F(20),R(4,20),X(20),X1(20),A(100,20),T(100)

20 CLS:PRINT"SIMULTANEOUS LINEAR DIFF EQUATIONS"

30 INPUT"NO. OF EQUATIONS=";D

40 INPUT"TIME INTERVAL, STEP SIZE=";T,DT

50 FOR I=1 TO D

60 PRINT"INITIAL VALUE OF X(";I;")=";

62 INPUT X1(I)

70 NEXT I
```

```
80 CLS

90 INPUT"PRINT INTERVAL=";NP

95 CLS:PRINT"****PROGRAM RUNNING****"

100 N1=N1+1

105 GOSUB 6000' CALC ANY CONSTANTS BEFORE START OF INTEGRATION

110 FOR C=1 TO D

120 X(C)=X1(C)

130 GOSUB 5000

140 R(1,C)=DT*F(C)

150 NEXT C

160 FOR C=1 TO D

170 X(C)=X1(C)+.5*R(1,C):X(0)=Y+.5*DT+XT

180 GOSUB 5000

190 R(2,C)=DT*F(C)

200 NEXT C

210 FOR C=1 TO D

220 X(C)=X1(C)+.5*R(2,C):X(0)=Y+.5*DT+XT

230 GOSUB 5000

240 R(3,C)=DT*F(C)

250 NEXT C

260 FOR C=1 TO D

270 X(C)=X1(C)+R(3,C):X(0)=Y+DT+XT

280 GOSUB 5000

290 R(4,C)=DT*F(C)

300 NEXT C

310 FOR C=1 TO D

320 X1(C)=X1(C)+(R(1,C)+2*R(2,C)+2*R(3,C)+R(4,C))/6

330 NEXT C
```

```
 340 XT=XT+DT

 350 IF N1=NP THEN GOSUB 1000

 360 IF XT>T THEN 380

 370 GOTO 100

 380 GOSUB 2000

 390 CLS:PRINT"PROGRAM EXECUTION TERMINATED"

 400 END

1000 REM SAVE ANSWERS FOR PRINTING

1010 N=N+1

1012 T(N)=XT

1020 FOR I=1 TO D

1030 A(N,I)=X1(I)

1040 NEXT I

1050 N1=0

1060 RETURN

2000 REM-PRINT ANSWERS

2005 CLS:PRINT"CALCULATIONS COMPLETED.....PRINTING ANSWERS"

2010 FOR I=1 TO N

2020 LPRINT "T=";T(I)

2030 LPRINT"------------------"

2040 FOR J=1 TO D

2050 LPRINT "X(";J;")=";A(I,J)

2060 NEXT J

2070 LPRINT"------------------"

2080 NEXT I

2090 RETURN

4999 REM-ADD EQUATIONS HERE......USE FORM F(1)=A*X(1)...ETC.

5000 K1=1.00:K2=.100
```

```
5010 F(1)=-K1*X(1)*X(2)

5020 F(2)=F(1)-K2*X(2)*X(3)

5030 F(3)=-F(1)-K2*X(2)*X(3)

5040 F(4)=K2*X(2)*X(3)

5050 RETURN

6000 REM-INSERT EXPRESSIONS FOR CALCULATED CONSTANTS HERE

6999 RETURN
```

NONLINEAR SIMULTANEOUS EQUATIONS

NEWTON is a program used to solve a single nonlinear equation or a system of nonlinear equations. The program is capable of solving up to 20 equations. It solves the equations using a Newton Raphson method in which the partial derivatives are calculated numerically. The best way to explain the program is by a demonstration.

PROBLEM: Solve the two equations defined as

$$F1 = X^2 + Y^2 - 3$$
$$F2 = 2X + Y$$

SOLUTION: Starting at line 1010, enter the equations as follows:

```
1010 F(1)=X(1)[2-X(2)[2-3
1020 F(2)=2*X(1)+X(2)
```

Now type RUN and enter the following data:

1. Number of equations
2. Initial guesses
3. Damping factor
4. Maximum number of iterations
5. Error tolerance

After 23 iterations, the answers are

$$X = -.776155$$
$$Y = 1.55229$$

EXAMPLE

READY

RUN

SIMULTANEOUS NONLINEAR EQUATIONS

TYPE THE NUMBER OF EQUATIONS? **2**

INPUT THE MAX NUMBER OF ITERATIONS? **100**

TYPE DAMPING FACTOR 0–1.0? **.5**

TYPE ERROR TOLERANCE (.001 – .0005)? **.0001**

TYPE THE INITIAL GUESS FOR EACH VARIABLE

X1 = ? **−1**

X2 = ? **1**

SOLUTION TO SYSTEM

X1 = −.776175   F1 = .0120425

X2 = 1.55229   F2 = −6.27041E−05

PROGRAM EXECUTION TERMINATED

READY

Nomenclature for NEWTON

NEWTON solves systems of nonlinear equations by filling a Jacobian matrix with the values of the partial derivatives at X_k and Y_k, inverting the matrix and multiplying times the function vector matrix. The solution to this linear system of equations is ΔX_k and ΔY_k. The next iteration will incorporate improved guesses for X_k and Y_k, namely $X_{k+1} = X_k + \Delta X_k$ and $Y_{k+1} = Y_k + \Delta Y_k$. The mathematical notation for a two-dimensional system would be:

$$f(x,y) = 0$$
$$g(x,y) = 0$$

Perry, R., *Chemical Engineer's Handbook,* McGraw Hill, 5th Ed, McGraw-Hill, 1973, New York, N.Y., pp. 254–255.

$$\Delta X_K \frac{\delta f}{\delta x} (X_k, Y_k) + \Delta Y_k \frac{\delta f}{\delta y} (X_k, Y_k) = -f(X_k, Y_k)$$

$$\Delta X_K \frac{\delta g}{\delta x} (X_k, Y_k) + \Delta Y_k \frac{\delta g}{\delta y} (X_k, Y_k) = -g(X_k, Y_k)$$

This is a system of linear equations of the form:

$$a_1 x_1 + b_1 x_2 = C_1$$
$$a_2 x_1 + b_2 x_2 = C_2$$

The equations are solved for ΔX_k and ΔY_k and added to the current value of X_k, Y_k to get the next approximation X_{k+}, Y_{k+1}. The procedure is repeated until the difference between the current and previous iteration over all variables is less than the desired error tolerances.

NEWTON calculates the partial derivatives numerically and solves the system of linear equations using a modified Gauss-Jordon elimination method, avoiding time consuming matrix inversion procedures.

```
    LISTING FOR BASIC PROGRAM "NEWTON"

 5 DIM X(20),F(20),A(20,21),F1(20),F2(20)

10 PRINT"SIMULTANEOUS NONLINEAR EQUATIONS"

20 INPUT"TYPE  THE NUMBER OF EQUATIONS";N

30 INPUT"INPUT THE MAX NUMBER OF ITERATIONS";N1

35 INPUT"TYPE  DAMPING FACTOR 0-1.0";D2

37 INPUT "TYPE  ERROR TOLERANCE (.0001-.0005)";E1

40 CLS:C=0

50 PRINT"TYPE  THE INITIAL GUESS FOR EACH VARIABLE"

60 FOR I=1 TO N

70 PRINT"X";I;"=";

80 INPUT X(I)

90 NEXT I

100 CLS
```

```
105 REM-FILL JACOBIAN MATRIX WITH PARTIAL DERVIATIVES

110 FOR C1=1 TO N1

115 TEMP=C

120 FOR I=1 TO N

130 D1=.001*X(I)

140 X(I)=X(I)+D1

150 FOR K=1 TO N

160 GOSUB 1000

170 F1(K)=F(K)

180 NEXT K

190 X(I)=X(I)+2*D1

200 FOR K=1 TO N

210 GOSUB 1000

220 F2(K)=F(K)

230 A(K,I)=(F2(K)-F1(K))/(2*D1)

240 NEXT K

250 X(I)=X(I)-D1

260 NEXT I

270 GOSUB 1000

275 REM-SOLUTION TO SYSTEM OF LINEAR EQUATIONS FROM

276 REM-"SOME COMMON BASIC PROGRAMS, 3RD EDITION", LON POOLE, MARY BOR
      CHERS, WITH PERMISSION  OSBORNE-MC GRAW HILL BOOKS @1979

280 FOR I=1 TO N

290 A(I,N+1)=-F(I)

300 NEXT I

310 FOR J=1 TO N

320 FOR I=J TO N

330  IF A(I,J)<>0 THEN 370
```

```
340 NEXT I

350 PRINT "NO UNIQUE SOLUTION"

360 GOTO 670

370 FOR K=1 TO N+1

380 X=A(J,K)

390 A(J,K)=A(I,K)

400 A(I,K)=X

410 NEXT K

420 Y=1/A(J,J)

430 FOR K=1 TO N+1

440 A(J,K)=Y*A(J,K)

450 NEXT K

460 FOR I=1 TO N

470 IF I=J THEN 520

480 Y=-A(I,J)

490 FOR K=1 TO N+1

500 A(I,K)=A(I,K)+Y*A(J,K)

510 NEXT K

520 NEXT I

530 NEXT J

540 FOR I=1 TO N

550 X(I)=X(I)+D2*A(I,N+1)

560 NEXT I

565 C=0

570 FOR I=1 TO N

580 C=C+X(I)

590 NEXT I

600 IF ABS((TEMP-C)/C)<=E1 THEN 630
```

```
610 NEXT C1

615 GOSUB 1000

620 PRINT "SYSTEM DID NOT CONVERGE":GOTO 640

630 CLS:PRINT"SOLUTION TO SYSTEM"

635 GOSUB 1000

640 FOR I=1 TO N

650 PRINT "X";I;"=";X(I);TAB(20);"F";I;"=";F(I)

660 NEXT I

670 PRINT"PROGRAM EXECUTION TERMINATED"

680 END

1000 REM-TYPE  SYSTEM OF EQUATIONS HERE

1001 REM-USE FORMAT F(1)=A*X(1)+B*X(2)[3+ETC.

1010 F(1)=X(1)[2-X(2)[2-3

1020 F(2)=2*X(1)+X(2)

1050 RETURN
```

2
HEAT TRANSFER

SHELL AND TUBE HEAT EXCHANGERS

EXCHANGE is a program for rating shell and tube heat exchangers where no change of phase takes place. The program is self-documenting and allows you to change parameters easily and recalculate the results based on the new information. The program calculates heat transfer coefficients, pressure drops, and temperature approach factors for multipass and multishelled heat exchangers. The calculated heat transfer area is based on the outside diameter (OD) of the tubing. The shell side pressure drops, and heat transfer coefficients assume segmental baffles with a 25% baffle cut. The information provided by EXCHANGE tends to be on the conserative side, so safety factors of 10–15% appear to be adequate.

Equations for EXCHANGE

Tube Side

$$A1 = [\pi(ID/12)^2 * NT/4]/N$$
$$G1 = W1/A1$$
$$R1 = (ID/12) * G1/(V1 * 2.42)$$
$$P1 = C1 * V1 * 2.42/K1$$
$$H1 = .027K1 * 12/ID * R1^{.8} * P1^{.333}$$
$$F1 = .048/(144 * R1^{.2})$$
$$VT = W1/(D1 * A1 * 3600)$$
$$\Delta P_1 = F1 * L * N(G1/3600)^2/(64.4 * D1 * ID/12)$$
$$\Delta P_2 = 4 * N * VT^2 * D1/(64.4 * 144)$$
$$\Delta P_T = [\Delta P_1 + \Delta P_2] * NS$$

Kern, D., *Process Heat Transfer,* 1st Ed., McGraw Hill, New York, 1950, pp. 221–251.

Shell Side

Square pitch: $DE = (4/P^2 - \pi * OD^2/4)/(\pi * OD * 12)$
Triangular pitch: $DE = (.43 * P^2 - .5\pi * OD^2/4) * 4/(\pi * OD * 12/2)$
$C = P\text{-}OD$
$AS = DS * C * B/(P * 144)$
$GS = W2/AS$
$VS = W2/(AS * D2 * 3600)$
$R2 = DE * GS/(V2 * 2.42)$
$P2 = C2 * VS * 2.42/K2$
$H2 = .36K2/DE * R2^{.55} * P2^{.333}$
If $R2 > 500$, $F2 = .009287 * R2^{-.15977}$
If $R2 > 500$, $F2 = .77338 * R2^{-.8774}$
$PS = [F2 * (GS/3600)^2 * DS/12 * (NB + L)] * NS/(64.4 * DE * D2)$

Heat Transfer Coefficient

$U = 1/(1/H1 + 1/H2 + FT + FS)$

LMTD

$LMTD = ABS \{[(T3 - T2) - (T4 - T1)]/\ln[(T3 - T2)/(T4 - T1)]\}$

Approach Factor

$S = (T2 - T1)/(T3 - T1)$
$R = (T3 - T4)/(T2 - T1)$
$X = [(R * S - 1)/(S - 1)]^{(1/NS)}$
$PX = (1 - X)/(R - X)$
$X1 = sqr(R^2 + 1)$
$X2 = \ln [(1 - PX)/(1 - R * PX)]$
$X3 = 2/PX - 1 - R + X1$
$X4 = 2/PX - 1 - R - X1$
$X5 = \ln(X3/X4)$
$FC = X1 * X2/[(R - 1) * X5]$

Heat Load

$Q = W1 * C1 * ABS(T1 - T2)$

Heat Transferred

$Q1 = A2 * U * LMTD * FC$
$A2 = OD/12 * \pi * NT * NS * L$

Nomenclature for EXCHANGE

VARIABLE	*DESCRIPTION*
Tube Side Variables	
T1	Inlet temp (°F)
T2	Outlet temp (°F)
ID	Tube ID (inside diameter) (in.)
OD	Tube OD (in.)
NT	Tube count
N	Number of tube passes
L	Tube length (ft)
W1	Mass flow rate (pph)
A1	Flow area (ft^2)
G1	Superficial flow rate (lb/hr ft^2)
R1	Tube Reynolds number
P1	Tube Prandtl number
H1	Tube heat transfer coefficient (Btu/hr ft^2 °F)
F1	Tube friction factor
VT	Tube velocity (ft/sec)
ΔP_1	Tube pressure drop (psi)
ΔP_2	Return losses (psi)
P	Tube pitch (in.)
D1	Tube fluid density (lb/ft^3)
K1	Tube fluid thermal conductivity (Btu/ft-hr-°F)
C1	Specific heat (Btu-/lb-°F)
FT	Tubeside fouling factor ($1/h_f$)
h_f	Fouling Factor (Btu/hr-ft^2-°F)
π	3.1416

VARIABLE	*DESCRIPTION*
Shell Side Variables	
T3	Inlet temperature (°F)
T4	Outlet temperature (°F)
W2	Shell mass flow rate (pph) (pounds per hour)
DS	Shell ID (in.)
C	Tube clearance (in.)
B	Baffles spacing (in.)
NB	No. of baffles
DE	Shell equivalent DIA (ft)
AS	Shell flow area (ft^2)
GS	Shell superficial flow rate (lb/hr-ft^2)

R2	Shell Reynolds number
P2	Shell PRANDTL number
H2	Shell heat transfer coeff (Btu/hr-ft^2-°F)
F2	Shell side friction factor
PS	Shell side pressure drop (psi)
NS	Number of shells per unit
D2	Shell fluid density (lb/ft^3)
K2	Shell fluid thermal conductivity (Btu/ft-hr-°F)
C2	Shell fluid specific heat (Btu/lb-°/F)
FS	Shell side fouling factor 1/h$_f$
h$_f$	Fouling factor (Btu/hr-ft^2-°F)
π	3.1416
U	Overall heat transfer coefficient (Btu/hr-ft^2-°F)
LMTD	Log mean temp difference (°F)
S	Variables for approach factor
R	Variables for approach factor
X	Variables for approach factor
PX	Variables for approach factor
X1	Variables for approach factor
X2	Variables for approach factor
X3	Variables for approach factor
X4	Variables for approach factor
X5	Variables for approach factor
FC	Approach factor
Q	Heat load on unit (Btu/hr)
A2	Outside area of unit (ft^2)

EXAMPLE

RUN

SHELL AND TUBE HEAT EXCHANGERS

TYPE OF EXCHANGERS EVALUATED:

1. COUNTER CURRENT-FORCED CONVECTION

2. NO CHANGE OF PHASE

TYPE IN PROBLEM TITLE? **TEST**

NO. TUBES=0

TUBE ID/OD: 0/0

PITCH(IN)=0

NO. TUBES, ID, OD, PITCH?   102,.680,.75,1.00

PITCH = TRIANGULAR
SQUARE(S) OR TRIANGULAR(T) PITCH?   T

TUBE LENGTH (FT) = 0
NUMBER OF PASSES = 0
TUBE LENGTH, NO. OF PASSES?   10,2

TUBE FLUID:
FOULING FACTOR = 0
NAME OF FLUID, FOULING FACTOR?   WATER, .0005

TEMP IN(F) = 0
TEMP OUT(F) = 0
SPECIFIC HEAT (BTU/LB-F) = 0
TEMP IN, TEMP OUT, SPECIFIC HEAT?   85,100,1

VISCOSITY(CP) = 0
DENSITY (LB/FT3) = 0
VISCOSITY, DENSITY?   1,62.3

THERMAL COND.(BTU/FT-HR-F) = 0
THERMAL COND.?   .345

MASS FLOW (LB/HR) = 0
MASS FLOW?   100000

SHELL PASSES = 0
SHELL ID (INCHES) = 0
NO. OF SHELL PASSES, SHELL ID?   1,12

NO. OF BAFFLES = 0
BAFFLE SPACING (INCHES) = 0
NO. OF BAFFLES, BAFFLE SPACING?   30,4

SHELL FLUID:

FOULING FACTOR=0

NAME OF FLUID, FOULING FACTOR?   **OIL,.001**

TEMP IN(F)=0

TEMP OUT(F)=0

SPECIFIC HEAT (BTU/LB-F)=0

TEMP IN, TEMP OUT, SPECIFIC HEAT?   **300,150,.5**

VISCOSITY(CP)=0

DENSITY(LB/FT3)=0

VISCOSITY, DENSITY?   **2.50,45.0**

THERMAL COND.(BTU/FT-HR-F)=0

THERMAL COND.?   **.065**

MASS FLOW RATE (LB/HR)=0

MASS FLOW RATE?   **20000**

MAKE ANY CHANGES (Y/N)?   **N**

SYSTEM:TEST

AREA OF UNIT=200.255   SHELLS/UNIT: 1

TUBES/UNIT=102   PASSES/SHELL=2

TUBE ID/OD .68/.75   PITCH=1 (T)

	TUBE SIDE	SHELL SIDE
FLUID	WATER	OIL
LBS/HR	100000	20000
THERMAL COND.	.345	.065
SPECIFIC HEAT	1	.5
DENSITY	62.3	45
VISCOSITY (CP)	1	2.5
TEMP IN(F)	85	300

TEMP OUT(F)	100	150
VELOCITY (FT/SEC)	3.46689	1.48148
PRESS. DROP (PSI)	.838282	2.1591
?		
PRANDTL NO.	7.01449	46.5385
REYNOLDS NO.	18207.2	2347.56
h-(Btu/hr-ft2-F)	804.876	101.466
FOULING FACTOR	5E-04	1E-03

OVERALL COEFFICIENT = 79.3779   LMTD = 120.114

APPROACH FACTOR = .971638   CORRECTED LMTD = 116.708

HEAT LOAD = 1.5E + 06   HEAT TRANSFERRED =

                                                         1.85537E + 06

SHELL DIA(IN) = 12   BAFFLE SPACING(IN) = 4

TUBE LENGTH(FT) = 10   NO. OF BAFFLES = 30

?

DO YOU WISH TO MAKE ANY CHANGES (Y/N)?   N

PROGRAM EXECUTION TERMINATED

READY

     LISTING FOR BASIC PROGRAM "EXCHANGE"

```
10 CLS:PI=3.1416

20 PRINT"SHELL AND TUBE HEAT EXCHANGERS"

30 PRINT"TYPE OF EXCHANGERS EVALUATED:"

40 PRINT"1. COUNTER CURRENT-FORCED CONVECTION"

50 PRINT"2. NO CHANGE OF PHASE"

60 INPUT"TYPE  PROBLEM TITLE";TITLE$

70 GOSUB 700
```

```
80 GOSUB 760

90 GOSUB 790

100 GOSUB 820

110 GOSUB 850

120 GOSUB 890

130 GOSUB 930

140 GOSUB 960

150 GOSUB 990

160 GOSUB 1020

170 GOSUB 1050

180 GOSUB 1080

190 GOSUB 1120

200 GOSUB 1150

210 GOSUB 1180

220 CLS

230 INPUT"MAKE ANY CHANGES(Y/N)";A$

240 IF A$="Y" THEN GOSUB 1540

250 A1=(ID/12)[2*PI/4*NT/N

260 G1=W1/A1

270 R1=(ID/12)*G1/(V1*2.42)

280 P1=C1*2.42*V1/K1

290 H1=.027*K1*12/ID*(R1)[.8*P1[.333

300 F1=.048/(144*R1[.2)

310 VT=W1/(D1*A1*3600)

320 X=F1*L*N*(G1/3600)[2/(64.4*D1*ID/12)

330 Y=4*N*VT[2*D1/(64.4*144)

340 PT=(X+Y)*NS

350 IF P$="S" THEN 380
```

```
360 DE=(.43*P[2-.5*PI*OD[2/4)*4/(PI*OD*12/2)

370 GOTO 400

380 DE=4*(P[2-PI*OD[2/4)/(PI*OD*12)

390 REM

400 C=P-OD

410 AS=DS*C*B/(P*144)

420 GS=W2/AS:VS=W2/(AS*D2*3600)

430 P2=C2*2.42*V2/K2

440 R2=DE*GS/(V2*2.42)

450 H2=.36*K2/DE*R2[.55*P2[.333

460 IF R2>500 THEN 490

470 F2=.77338*R2[-.8774

480 GOTO 500

490 F2=.009287*R2[-.15977

500 REM

510 PS=(F2*(GS/3600)[2*DS/12*(NB+1))*NS/(64.4*DE*D2)

520 U=1/(1/H1+1/H2+FT+FS)

530 IF N=1 THEN 650

540 S=(T2-T1)/(T3-T1)

550 R=(T3-T4)/(T2-T1)

560 IF R=1 THEN R=1.00001

570 X=((R*S-1)/(S-1))[(1/NS)

580 PX=(1-X)/(R-X)

590 X1=SQR(R[2+1)

600 X2=LOG((1-PX)/(1-R*PX))

610 X3=2/PX-1-R+X1

620 X4=2/PX-1-R-X1

630 X5=LOG(X3/X4)
```

```
640 FC=X1*X2/((R-1)*X5)

650 LMTD=ABS(((T3-T2)-(T4-T1))/LOG((T3-T2)/(T4-T1)))

660 IF N=1 THEN FC=1

670 Q=W1*C1*ABS(T1-T2)

680 A2=OD/12*PI*NT*NS*L

690 GOTO 1210

700 CLS

710 PRINT"NO. TUBES=";NT

720 PRINT"TUBE ID/OD=";ID;"/";OD

730 PRINT"PITCH(IN)=";P

740 INPUT"NO. TUBES, ID, OD, PITCH";NT,ID,OD,P

750 CLS:RETURN

760 CLS:IF P$="S" THEN PRINT"PITCH=SQUARE"ELSE PRINT"PITCH=TRIANGULAR"

770 INPUT"SQUARE(S) OR TRIANGULAR(T) PITCH";P$

780 CLS:RETURN

790 CLS:PRINT"TUBE LENGTH(FT)=";L:PRINT"NUMBER OF PASSES=";N

800 INPUT"TUBE LENGTH, NO. OF PASSES";L,N

810 CLS:RETURN

820 CLS:PRINT"TUBE FLUID=";T$:PRINT"FOULING FACTOR=";FT

830 INPUT"NAME OF FLUID, FOULING FACTOR";T$,FT

840 CLS:RETURN

850 CLS:PRINT"TEMP IN(F)=";T1:PRINT"TEMP OUT(F)=";T2

860 PRINT"SPECIFIC HEAT(BTU/LB-F)=";C1

870 INPUT"TEMP IN, TEMP OUT, SPECIFIC HEAT";T1,T2,C1

880 CLS:RETURN

890 CLS:PRINT"VISCOSITY(CP)=";V1

900 PRINT"DENSITY (LB/FT3)=";D1

910 INPUT"VISCOSITY, DENSITY";V1,D1
```

```
 920 CLS:RETURN

 930 CLS:PRINT"THERMAL COND.(BTU/FT-HR-F)=";K1

 940 INPUT"THERMAL COND.";K1

 950 CLS:RETURN

 960 CLS:PRINT"MASS FLOW (LB/HR)=";W1

 970 INPUT"MASS FLOW";W1

 980 CLS:RETURN

 990 CLS:PRINT"SHELL PASSES=";NS:PRINT"SHELL ID(INCHES)=";DS

1000 INPUT"NO. OF SHELL PASSES, SHELL ID";NS,DS

1010 CLS:RETURN

1020 CLS:PRINT"NO. OF BAFFLES=";NB:PRINT"BAFFLE SPACING (INCHES)=";B

1030 INPUT"NO. OF BAFFLES, BAFFLE SPACING";NB,B

1040 CLS:RETURN

1050 CLS:PRINT"SHELL FLUID:";S$:PRINT"FOULING FACTOR=";FS

1060 INPUT"NAME OF FLUID, FOULING FACTOR";S$,FS

1070 CLS:RETURN

1080 CLS:PRINT"TEMP IN(F)=";T3:PRINT"TEMP OUT(F)=";T4

1090 PRINT"SPECIFIC HEAT(BTU/LB-F)=";C2

1100 INPUT" TEMP IN, TEMP OUT, SPECIFIC HEAT";T3,T4,C2

1110 CLS:RETURN

1120 CLS:PRINT"VISCOSITY(CP)=";V2:PRINT"DENSITY(LB/FT3)=";D2

1130 INPUT"VISCOSITY, DENSITY";V2,D2

1140 CLS:RETURN

1150 CLS:PRINT"THERMAL COND.(BTU/FT-HR-F)=";K2

1160 INPUT"THERMAL COND.";K2

1170 CLS:RETURN

1180 CLS:PRINT"MASS FLOW RATE (LB/HR)=";W2

1190 INPUT"MASS FLOW RATE";W2
```

```
1200 CLS:RETURN

1210 X$=" "

1220 PRINT"SYSTEM:";TITLE$

1230 PRINT"AREA OF UNIT=";A2;TAB(30);"SHELLS/UNIT:";NS

1240 PRINT"TUBES/UNIT=";NT*NS;TAB(30);"PASSES/SHELL=";N

1250 PRINT"TUBE ID/OD";ID;"/";OD;TAB(30);"PITCH=";P;"(";P$;")"

1260 PRINT TAB(20);"TUBE SIDE";TAB(40);"SHELL SIDE"

1270 PRINT"FLUID";TAB(20);T$;TAB(40);S$

1280 PRINT"LBS/HR";TAB(20);W1;TAB(40);W2

1290 PRINT"THERMAL COND.";TAB(20);K1;TAB(40);K2

1300 PRINT"SPECIFIC HEAT";TAB(20);C1;TAB(40);C2

1310 PRINT"DENSITY";TAB(20);D1;TAB(40);D2

1320 PRINT"VISCOSITY(CP)";TAB(20);V1;TAB(40);V2

1330 PRINT"TEMP IN(F)";TAB(20);T1;TAB(40);T3

1340 PRINT"TEMP OUT(F)";TAB(20);T2;TAB(40);T4

1350 IF X$="Y" THEN RETURN

1360 PRINT"VELOCITY(FT/SEC)";TAB(20);VT;TAB(40);VS

1370 PRINT"PRESS. DROP (PSI)";TAB(20);PT;TAB(40);PS

1380 INPUT Z$

1390 CLS

1395 PRINT"PRANDTL NO.";TAB(20);P1;TAB(40);P2

1400 PRINT"REYNOLDS NO.";TAB(20);R1;TAB(40);R2

1410 PRINT"h-(Btu/hr-ft2-F)";TAB(20);H1;TAB(40);H2

1420 PRINT"FOULING FACTOR";TAB(20);FT;TAB(40);FS

1430 PRINT

1440 PRINT"OVERALL COEFFICENT=";U;TAB(30);"LMTD=";LMTD

1450 PRINT"APPROACH FACTOR=";FC;TAB(30);"CORRECTED LMTD=";LMTD*FC

1460 PRINT"HEAT LOAD=";Q;TAB(30);"HEAT TRANSFERRED=";U*A2*LMTD*FC
```

```
1470 PRINT"SHELL DIA(IN)=";DS;TAB(30);"BAFFLE SPACING(IN)=";B

1480 PRINT"TUBE LENGTH(FT)=";L;TAB(30);"NO. OF BAFFLES=";NB

1490 INPUT Z$:CLS

1500 INPUT"DO YOU WISH TC MAKE ANY CHANGES (Y/N)?";A$

1510 IF A$="Y" THEN GOSUB 1540 :GOTO 250

1520 PRINT"PROGRAM EXECUTION TERMINATED"

1530 END

1540 CLS:INPUT"CHANGE SHELL SIDE(1) OR TUBE SIDE(2),OR 0 TO END";Z

1550 IF Z=0 THEN RETURN

1560 ON Z GOSUB 1580 ,1730

1570 GOTO 1540

1580 CLS

1590 PRINT"SHELLSIDE DATA"

1600 PRINT

1610 PRINT"1. NO. OF SHELL PASSES,SHELL ID"

1620 PRINT"2. NO. OF BAFFLES, BAFFLE SPACING"

1630 PRINT"3. NAME OF FLUID, FOULING FACTOR"

1640 PRINT"4. TEMP(IN),TEMP(OUT),SPECIFIC HEAT

1650 PRINT"5. VISCOSTIY, DENSITY"

1660 PRINT"6. THERMAL CONDUCTIVITY"

1670 PRINT"7. MASS FLOW RATE"

1680 PRINT

1690 INPUT"CHANGE WHICH VARIABLE(0-7), 0 TO RETURN";Z1

1700 IF Z1=0 THEN RETURN

1710 ON Z1 GOSUB 990   ,1020 ,1050 ,1080 ,1120 ,1150 ,1180

1720 GOTO 1580

1730 CLS

1740 PRINT"TUBE SIDE DATA"
```

```
1750 PRINT

1760 PRINT"1. NO. OF TUBES PER SHELL, ID, OD, PITCH"

1770 PRINT"2. SQUARE OR TRIANGULAR PITCH"

1780 PRINT"3. TUBE LENGTH, NO. OF PASSES"

1790 PRINT"4. FLUID NAME, FOULING FACTOR"

1800 PRINT"5. TEMP(IN),TEMP(OUT), SPECIFIC HEAT"

1810 PRINT"6. VISCOSITY, DENSITY"

1820 PRINT"7. THERMAL CONDUCTIVITY"

1830 PRINT"8. MASS FLOW RATE"

1840 PRINT

1850 INPUT"CHANGE WHICH VARIABLE(0-8),0 TO END";Z3

1860 IF Z3=0 THEN RETURN

1870 ON Z3 GOSUB 700  ,760  ,790  ,820  ,850  ,890  ,930  ,960

1880 GOTO 1730
```

DOUBLE PIPE HEAT EXCHANGERS

Double pipe heat exchangers are very common pieces of equipment in the chemical processing industry. Although they are very simple mechanically, they can be very difficult to rate with just a programmable calculator. The program DBLPIPE will rate double pipe heat exchangers with and without longitudal fins. The method used by the program is that outlined by Kern.[1] The overall heat transfer coefficients are referenced to the *inside area* of the tube. The heat transferred by the unit is given by

$$Q = U_i A_i LMTD$$

When you specify STEAM on the tube side, the program will prompt you for a condensing coefficient and does not calculate a tube side pressure drop.

When fins are used and the A_f/A_i ratio is 2.5–3:1, caution must be exercised when assigning fouling factors. If the expected fouling factor in a shell and tube unit is, for example, .002, divide this fouling factor by the

---

[1]Kern, D., *Process Heat Transfer,* 1st Ed., McGraw Hill, New York, NY, 1950, pp. 515–530.

A_f/A_i ratio to get an effective shell side fouling factor for a finned unit. The program is highly structured to allow you to program in preferred heat transfer and pressure drop relationships.

When the program was tested against published examples, the results were generally on the conservative side (i.e., lower U_i). Safety factors of 10–15% appear adequate.

Equations for DBLPIPE

Tube Side

$$A1 = (D1/12)^2 * \pi/4$$
$$GT = WT/A1$$
$$VT = (WT/R1H0)/(A1 * 3600)$$
$$RT = (D1/12) * GT/(V1 * 2.42)$$
$$PT = C1P * V1 * 2.42/K1$$
$$HT = K1 * 12/D1 * .023 * RT^{.80} * PT^{.333}$$
$$F1 = 0.00333/RT^{.20}$$
If RT < 2100, $F1 = RT/16$
$$DTP = 4 * F1 * L * NP * (GT/3600)^2/(64.4 * R1H0 * D1/12)$$

Shell Side

$$WP = (D2 * \pi - NF * B + NF * H * 2)/12$$
$$A3 = [(D2/12)^2 * \pi/4] + NF * H * B/144$$
$$A4 = (D3/12)^2 * \pi/4$$
$$FLWA = A4 - A3$$
$$DE = 4 * FLWA/WP$$
$$GS = WS/FLWA$$
$$VS = (WS/R2H0)/(FLWA * 3600)$$
$$RS = DE * GS/(2.42 * V2)$$
$$PS = C2P * 2.42 * V2/K2$$
If RS < 2100, $HS = 1.68 * K2/DE * RS^{.333} * PS^{.333} * (DE/L)^{.333}$
If RS > 2100, $HS = K2/DE * .023 * RS^{.8} * PS^{.333}$
$$X1 = H/12$$
$$B1 = B/24$$
$$N = sqr[(HS * X1^2)/(K3 * B1)]$$
$$THN = [-exp(-N)]/[exp(N) + exp(-N)] * 2 + 1$$
$$EFF = THN/N$$
$$AF = NF * 2 * H/12$$
$$AO = (D2 * \pi - NF * B1)/12$$
$$AI = D1/12 * \pi$$
$$HFI = HS * (EFF * AF + AO)/AI$$

Shell Side Pressure Drop

$DE = 4 * FLWA/(WP + D3 * \pi/12)$
$RE = DE * GS/(2.42 * V2)$
$F2 = .000333/RE^{.25}$
If $RE < 2100$, $F2 = 16/RE$
$DSP = 4 * F2 * L * NP * (GS/3600)^2/(64.4 * R2HO * DE)$
If $NF > 0$, $DSP = DSP * 2$

LMTD Equations

$X = ABS(T1 - T4)$ $LMTD = (X - Y)/LN(X/Y)$ COUNTER-
$Y = ABS(T2 - T3)$ CURRENT
$X = ABS(T1 - T2)$ $LMTD = (X - Y)/LN(X/Y)$ COCURRENT
$Y = ABS(T3 - T4)$

Heat Balances

$Q1 = WT * C1P * ABS(T1 - T2)$
$Q2 = WS * C2P * ABS(T3 - T4)$
$U = 1/(1/HT + 1/HFI + FT + FS * (AF + AO)/AI)$
$Q = U * L * NP * AI * LMTD$

Nomenclature for DBLPIPE

VARIABLE	*DESCRIPTION*
Tube Side	
T1	Inlet temperature (°F)
T2	Outlet temperature (°F)
D1	Tube ID (in.)
WT	Mass flow rate (pph)
C1P	Specific heat (Btu/lb-°F)
K1	Thermal conductivity (Btu/ft-hr-°F)
R1HO	Density (lb/ft³)
V1	Viscosity (cp)
A1	Tube flow area (ft²)
GT	Superficial mass flow rate (lb/hr-ft²)
VT	Tube velocity (ft/sec)
RT	Tube Reynolds Number
PT	Tube Prandlt Number
HT	Tube heat transfer coefficient (Btu/hr-ft²-°F)
F1	Friction factor

DTP	Press drop (psi)
L	Tube length (ft)
NP	Number of passes
FT	Fouling factor

Shell Side

T3	Inlet temperature (°F)
T4	Outlet temperature (°F)
D2	Tube OD (in.)
WS	Mass flow rate (pph)
C2P	Specific heat (Btu/LB − °F)
K2	Thermal conductivity (Btu/ft-hr-°F)
R2H0	Density (lb/ft^3)
V2	Viscosity (cp)
VS	Shell side velocity (ft/sec)
RS	Reynolds number
PS	PRANDTL number
F2	Friction factor
NF	Number of fins
H	Fin height (in.)
B	Fin thickness (in.)
A3	Temproary variable
A4	Temporary variable
FLWA	Flow area (ft^2)
DE	Hydraulic radius (ft)
HS	Shell heat transfer coefficient (Btu/hr-ft^2-°F)
X1	Temperature variable
B1	B/24 (ft)
N	Fin modulus
THN	Hyperbolic Tangent of N
K3	Fin metal thermal conductivity (Btu/ft-hr-°F)
EFF	Fin efficiency (arc)
AF	Fin area (ft^2/ft)
AO	OD tube area (ft^2/ft)
AI	Inside tube area (ft^2/ft)
HFI	Effective shell side heat transfer coefficient (Btu/hr-ft^2-°F)
DSP	Pressure drop (psi)
FS	Shell side fouling factor
LMTD	Log mean temperature difference (°F)

Heat Balances

Q1	Tube side heat load (Btu/hr)
Q2	Shell side Heat Load (Btu/hr)
Q	Heat transferred (Btu/hr)
U	Overall heat transfer coefficient (Btu/hr-ft^2-°F)
π	3.1416

EXAMPLE

RUN

DOUBLE PIPE HEAT EXCHANGERS

FORCED CONVECTION NO CHANGE OF PHASE

SYSTEM TITLE:? **TEST**

TUBE SIDE

FLUID NAME=

MASS FLOW RATE (PPH)=0

DENSITY (LB/FT3)=0

FOULING FACTOR=0

FLUID NAME=? **WATER**

MASS FLOW RATE (PPH)=? **10000**

DENSITY (LB/FT3)=? **62.3**

FOULING FACTOR=? **.001**

TUBE SIDE

TUBE ID(IN)=0

TUBE OD(IN)=0

TUBE ID(IN)=? **.970**

TUBE OD(IN)=? **1**

TUBE SIDE

TUBE LENGTH(FT)=0

NO. OF PASSES=0

TUBE LENGTH(FT)=?   12

NO. OF PASSES=?   5

TUBE SIDE

TEMP IN(F)=0

TEMP OUT(F)=0

TEMP IN(F)=?   80

TEMP OUT(F)=?   100

TUBE SIDE

VISCOSITY(CP)=0

SPECIFIC HEAT (BTU/LB-F)=0

THERMAL COND. (BTU/HR-FT-F)=0

VISCOSITY(CP)=?   1

SPECIFIC HEAT (BTU/LB-F)=?   1

THERMAL COND. (BTU/HR-FT-F)=?   .345

SHELL SIDE

FLUID NAME=

MASS FLOW RATE (PPH)=0

DENSITY (LB/FT3)=0

FOULING FACTOR=0

FLUID NAME=?   OIL

MASS FLOW RATE(PPH)=?   10000

DENSITY(LB/FT3)=?   45

FOULING FACTOR=?   .0005

SHELL SIDE

SHELL ID (IN)=0

SHELL ID (IN)=?   2.068

SHELL SIDE

TEMP IN(F)=0

TEMP OUT(F)=0

TEMP IN(F)=? 250

TEMP OUT(F)=? 210

SHELL SIDE

NO. OF FINS=0

FIN HT(IN)=0

FIN THICKNESS(IN)=0

METAL COND.(BTU/HR-FT-F)=0

NO. OF FINS=? 24

FIN HT(IN)=? 50

FIN THICKNESS (IN)=? .034

METAL COND.(BTU/HR-FT-F)=? 25

SHELL SIDE

VISCOSITY(CP)=0

SPECIFIC HEAT (BTU/LB-F)=0

THERMAL COND. (BTU/HR-FT-F)=0

VISCOSITY(CP)=? 2.5

SPECIFIC HEAT(BTU/HR-F)=? .50

THERMAL COND. (BTU/HR-FT-F)=? .065

SHELL SIDE

PIPING ARRANGEMENT=

COUNTERCURRENT(1) OR COCURRENT(2):? 1

MAKE ANY CHANGES (Y/N)? N

DESCRIPTION:TEST

TUBE LENGTH(FT)=12 NO. OF PASSES=5

TUBE AREA (FT²) = 15.2368   FIN AREA (FT2) = 48

FIN HEIGHT(IN) = .5   FIN THICKNESS (IN) = .034

NO. OF FINS = 24   TOTAL AREA (FT2) = 131.628

	TUBE SIDE	SHELL SIDE
FLUID	WATER	OIL
ID/OD(IN)	.97/1	2.068/--
FLOW RATE (PPH)	10000	10000
TEMP IN(F)	80	250
TEMP OUT(F)	100	210
SPECIFIC HEAT (BTU/LB-F)	1	.5
THERMAL COND. (BTU/HR-FT-F)	.345	.065
DENSITY (LB/FT3)	62.3	45
VISCOSITY (CP)	1	2.5
VELOCITY (FT/SEC)	8.68836	4.10485
Re	65088.4	3013.75
Pr	7.01449	46.5385
PRESSURE DROP (PSI)	7.8676	12.2045
FOULING FACTOR	1E-03	5E-04
h-(BTU/HR-FT2-F)	1331.91	118.921

FLOW ARRANGEMENT = COUNTERCURRENT

LMTD(F) = 139.762

OVERALL COEFFICIENT(BTU/HR-FT2-F) = 122.153

FIN EFFICIENCY(%) = 40.7606

HEAT LOAD (BTU/HR) = 200000

HEAT TRANSFERRED (BTU/HR) = 260126

MAKE CHANGES(Y/N):? **N**

PROGRAM EXECUTION TERMINATED

READY

```
    LISTING FOR BASIC PROGRAM "DBLPIPE"

  5 CLS
 10 PI=3.1416:T$="TUBE SIDE":S$="SHELL SIDE"
 20 PRINT"DOUBLE PIPE HEAT EXCHANGERS"
 30 PRINT"FORCED CONVECTION NO CHANGE OF PHASE"
 40 PRINT
 45 INPUT"SYSTEM TITLE:";D$
 50 REM-INPUT DATA
 60 GOSUB 2000
 70 GOSUB 3000
 80 GOSUB 4000
 90 GOSUB 5000
100 GOSUB 6000
110 GOSUB 7000
120 GOSUB 8000
125 GOSUB 8500
130 GOSUB 9000
140 GOSUB 10000
150 GOSUB 11000
152 CLS
154 INPUT"MAKE ANY CHANGES(Y/N)";A$
156 IF A$="Y" THEN GOSUB 20000
160 A$=" "
170 REM-START CALCULATIONS
```

```
175 IF NF=0 THEN B=0:H=0:EFF=0

180 GOSUB 12000'CALCULATE LMTD

190 GOSUB 13000'CALCULATE HEAT LOAD

200 GOSUB 14000'CALCULATE TUBE SIDE HEAT TRANSFER COEFFICIENT

210 GOSUB 15000'CALCULATE SHELL SIDE COEFFICIENT

215 GOSUB 15500'CALCULATE FIN EFFICIENCY

220 GOSUB 16000'CALCULATE TUBE SIDE PRESSURE DROP

230 GOSUB 17000'CALCULATE SHELL SIDE PRESSURE DROP

240 GOSUB 18000'CALCULATE ACTUAL HEAT TRANSFERRED

250 LPRINT"DESCRIPTION:";D$

260 LPRINT

270 LPRINT"TUBE LENGTH(FT)=";L,,"NO. OF PASSES=";NP

280 LPRINT"TUBE AREA(FT2)=";AI*L*NP,,"FIN AREA(FT2)=;"AF*L*NP

290 LPRINT"FIN HEIGHT(IN)=";H,,"FIN THICKNESS(IN)=";B

300 LPRINT"NO. OF FINS=";NF,,"TOTAL AREA(FT2)=";WP*L*NP

310 LPRINT

320 LPRINT,,T$,,S$

330 LPRINT"FLUID",,FT$,,FS$

340 LPRINT"ID/OD(IN)",,D1;"/";D2,,D3;"/--"

350 LPRINT"FLOW RATE(PPH)",,WT,,WS

360 LPRINT"TEMP IN(F)",,T1,,T3

370 LPRINT"TEMP OUT(F)",,T2,,T4

380 LPRINT"SPECIFIC HEAT(BTU/LB-F)",C1P,,C2P

390 LPRINT"THERMAL COND.(BTU/HR-FT-F)",K1,,K2

400 LPRINT"DENSITY(LB/FT3)",,R1HO,,R2HO

405 LPRINT"VISCOSITY(CP)",,V1,,V2

410 LPRINT"VELOCITY(FT/SEC)",VT,,VS

430 LPRINT"Re",,RT,,RS
```

```
440 LPRINT"Pr",,PT,,PS

450 LPRINT"PRESSURE DROP(PSI)",DTP,,DSP

460 LPRINT"FOULING FACTOR",,FT,,FS

470 LPRINT"h-(BTU/HR-FT2-F)",HT,,HS

480 LPRINT

485 LPRINT"FLOW ARRANGEMENT=";FLOW$

487 LPRINT"LMTD(F)=";LMTD

488 LPRINT"OVERALL COEFFICIENT(BTU/HR-FT2-F)=";U

490 LPRINT"FIN EFFICIENCY(%)=";EFF*100

500 LPRINT"HEAT LOAD(BTU/HR)=";Q1

510 LPRINT"HEAT TRANSFERRED(BTU/HR)=";Q

520 LPRINT:LPRINT:LPRINT

530 CLS

540 INPUT"MAKE CHANGES(Y/N):";A$

550 IF A$="Y" THEN GOSUB 20000:GOTO 170

560 PRINT"EXECUTION TERMINATED"

570 END

2000 CLS

2010 PRINT T$

2020 PRINT

2030 PRINT"FLUID NAME=";FT$

2040 PRINT"MASS FLOW RATE(PPH)=";WT

2050 PRINT"DENSITY(LB/FT3)=";R1HO

2055 PRINT"FOULING FACTOR=";FT

2060 PRINT

2070 INPUT"FLUID NAME=";FT$

2080 INPUT"MASS FLOW RATE(PPH)=";WT

2090 INPUT"DENSITY(LB/FT3)=";R1HO
```

```
2095 INPUT"FOULING FACTOR=";FT

2097 IF FT$="STEAM" THEN INPUT"CONDENSING COEFFICIENT=";HT

2100 RETURN

3000 CLS

3010 PRINT T$

3020 PRINT

3030 PRINT"TUBE ID(IN)=";D1

3040 PRINT"TUBE OD(IN)=";D2

3045 PRINT

3050 INPUT"TUBE ID(IN)=";D1

3060 INPUT"TUBE OD(IN)=";D2

3070 RETURN

4000 CLS

4010 PRINT T$

4020 PRINT

4030 PRINT"TUBE LENGTH(FT)=";L

4040 PRINT"NO. OF PASSES=";NP

4050 PRINT

4060 INPUT"TUBE LENGTH(FT)=";L

4070 INPUT"NO. OF PASSES=";NP

4080 RETURN

5000 CLS

5010 PRINT T$

5020 PRINT

5030 PRINT"TEMP IN(F)=";T1

5040 PRINT"TEMP OUT(F)=";T2

5050 PRINT

5060 INPUT"TEMP IN(F)=";T1
```

```
5070 INPUT"TEMP OUT(F)=";T2

5080 RETURN

6000 CLS

6005 IF FT$="STEAM" THEN RETURN

6010 PRINT T$

6020 PRINT

6030 PRINT"VISCOSITY(CP)=";V1

6040 PRINT"SPECIFIC HEAT (BTU/LB-F)=";C1P

6050 PRINT"THERMAL COND.(BTU/HR-FT-F)=";K1

6060 PRINT

6070 INPUT"VISCOSITY(CP)=";V1

6080 INPUT"SPECIFIC HEAT (BTU/LB-F)=";C1P

6090 INPUT"THERMAL COND.(BTU/HR-FT-F)=";K1

6100 RETURN

7000 CLS

7010 PRINT S$

7020 PRINT"FLUID NAME=";FS$

7030 PRINT"MASS FLOW RATE(PPH)=";WS

7040 PRINT"DENSITY(LB/FT3)=";RZHO

7045 PRINT"FOULING FACTOR=";FS

7050 PRINT

7060 INPUT"FLUID NAME=";FS$

7065 INPUT"MASS FLOW RATE(PPH)=";WS

7070 INPUT"DENSITY(LB/FT3)=";RZHO

7075 INPUT"FOULING FACTOR=";FS

7090 RETURN

8000 CLS

8010 PRINT S$
```

```
8020 PRINT
8030 PRINT"SHELL ID(IN)=";D3
8040 PRINT
8050 INPUT"SHELL ID(IN)=";D3
8060 RETURN
8500 CLS
8505 PRINT S$
8510 PRINT
8520 PRINT"TEMP IN(F)=";T3
8530 PRINT"TEMP OUT(F)=";T4
8540 PRINT
8550 INPUT"TEMP IN(F)=";T3
8560 INPUT"TEMP OUT(F)=";T4
8570 RETURN
9000 CLS
9010 PRINT S$
9020 PRINT
9030 PRINT"NO. OF FINS=";NF
9040 PRINT"FIN HT(IN)=";H
9050 PRINT"FIN THICKNESS(IN)=";B
9060 PRINT"METAL COND.(BTU/HR-FT-F)=";K3
9070 PRINT
9080 INPUT"NO. OF FINS=";NF
9090 INPUT"FIN HT(IN)=";H
9100 INPUT"FIN THICKNESS(IN)=";B
9110 INPUT"METAL COND.(BTU/HR-FT-F)=";K3
9120 RETURN
10000 CLS
```

```
10010 PRINT S$

10020 PRINT

10030 PRINT"VISCOSITY(CP)=";V2

10040 PRINT"SPECIFIC HEAT (BTU/LB-F)=";C2P

10050 PRINT"THERMAL COND.(BTU/HR-FT-F)=";K2

10060 PRINT

10070 INPUT"VISCOSITY(CP)=";V2

10080 INPUT"SPECIFIC HEAT(BTU/HR-F)=";C2P

10090 INPUT"THERMAL COND.(BTU/HR-FT-F)=";K2

10100 RETURN

11000 CLS

11010 PRINT S$

11020 PRINT

11030 PRINT"PIPING ARRANGEMENT=";P$

11040 PRINT

11050 INPUT"COUNTERCURRENT(1) OR COCURRENT(2):";P

11052 IF P=1 THEN FLOW$="COUNTERCURRENT"

11054 IF P=2 THEN FLOW$="COCURRENT"

11060 RETURN

12000 IF P=2 THEN 12050

12010 X=ABS(T1-T4)

12020 Y=ABS(T2-T3)

12030 LMTD=(X-Y)/LOG(X/Y)

12040 GOTO 12080

12050 X=ABS(T1-T2)

12060 Y=ABS(T3-T4)

12070 LMTD=(X-Y)/LOG(X/Y)

12080 RETURN
```

```
13000 Q1=WT*C1P*ABS(T1-T2)

13010 Q2=WS*C2P*ABS(T3-T4)

13015 IF FT$="STEAM" THEN Q1=Q2:RETURN

13020 IF ABS((Q1-Q2)/Q1) <=.05 THEN  13070

13030 PRINT"***WARNING***"

13040 PRINT"HEAT BALANCES DO NOT CHECK"

13050 INPUT"CONTINUE(Y/N)";A$

13060 IF A$="N" THEN GOSUB 5000:GOSUB 8500:GOTO 13070

13070 RETURN

14000 IF FT$="STEAM" THEN RETURN

14005 A1=(D1/12)[2/4*PI

14010 GT=WT/A1:VT=(WT/R1HO)/(A1*3600)

14020 RT=(D1/12)*GT/(2.42*V1)

14030 PT=C1P*2.42*V1/K1

14040 HT=K1*12/D1*.023*RT[.80*PT[.333

14050 RETURN

15000 WP=(D2*PI-NF*B+NF*H*2)/12

15010 A3=(D2/12)[2/4*PI+NF*H*B/144

15020 A4=(D3/12)[2/4*PI

15030 FLWA=A4-A3

15040 DE=4*FLWA/WP

15050 GS=WS/FLWA:VS=(WS/RZHO)/(FLWA*3600)

15060 RS=DE*GS/(2.42*V2)

15070 PS=C2P*2.42*V2/K2

15075 IF RS)2100 THEN 15080

15076 HS=1.68*K2/DE*RS[.333*PS[.333*(DE/L)[.333

15077 GOTO 15090

15080 HS=K2/DE*.023*RS[.8*PS[.333

15090 RETURN
```

```
15500 IF NF=0 THEN 15560

15510 X1=H/12

15520 B1=B/24

15530 N=SQR((HS*X1[2)/(K3*B1))

15540 TNH=(-EXP(-N))/(EXP(N)+EXP(-N))*2+1

15550 EFF=TNH/N

15560 AF=NF*2*H/12

15570 AO=(D2*PI-NF*B1)/12

15580 AI=D1/12*PI

15590 HFI=HS*(EFF*AF+AO)/AI

15600 RETURN

16000 IF FT$="STEAM" THEN RETURN

16002 F1=.000333/RT[.2

16005 IF RT<2100 THEN F1=16/(144*RT)

16010 DTP=4*F1*L*NP*(GT/3600)[2/(64.4*R1HO*D1/12)

16020 RETURN

16999 REM-CALC SHELL SIDE PRESSURE DROP

17000 DE=4*FLWA/(WP+D3*PI/12)

17010 RE=DE*GS/(2.42*VZ)

17020 F2=.000333/RE[.25

17030 IF RE<2100 THEN F2=16/(144*RE)

17040 DSP=4*F2*L*NP*(GS/3600)[2/(64.4*R2HO*DE)

17045 IF NF)0 THEN DSP=DSP*2

17050 RETURN

18000 REM-CALC HEAT TRANSFERRED

18010 X=1/HT+1/HFI+FT+FS*(AF+AO)/AI

18015 U=1/X

18020 Q=U*L*NP*AI*LMTD

18030 RETURN
```

```
20000 REM-EDITING ROUTINE

20010 GOSUB 21000

20020 INPUT"HOW MANY CHANGES, ENTER 0 TO END";N

20030 IF N=0 THEN RETURN

20040 CLS

20050 FOR I=1 TO N

20060 GOSUB 21000

20070 INPUT"CHANGE WHICH PARAMETER(0-10), ENTER 0 TO END";C

20075 IF C=0 THEN 20100

20080 ON C GOSUB 2000,3000,4000,5000,6000,7000,8000,8500,9000,10000,1100
      0

20090 NEXT I

20100 RETURN

21000 CLS

21010 PRINT T$

21020 PRINT"1.MASS FLOW, DENSITY, FOULING FACTOR"

21030 PRINT"2.TUBE ID/OD"

21040 PRINT"3.TUBE LENGTH, NO. OF PASSES"

21050 PRINT"4.TEMP IN/OUT"

21060 PRINT"5.VISCOSITY, SPECIFIC HEAT, THERMAL COND."

21080 PRINT S$

21090 PRINT"6.MASS FLOW, DENSITY, FOULING FACTOR"

21100 PRINT"7.SHELL ID"

21110 PRINT"8. TEMP IN/OUT

21120 PRINT"9.NO. OF FINS, HT, FIN THICKNESS, METAL COND."

21130 PRINT"10.VISCOSITY, SPECIFIC HEAT, THERMAL COND."

21140 PRINT"11.PIPING ARRANGEMENT"

21150 RETURN
```

3
THERMODYNAMICS

CHEMICAL REACTION STOICHIOMETRY

BALANCE is a general-purpose program designed to generate linearly independent chemical reactions that define the chemical equilibrium of a particular system. In general, if there are C components and M elements, the number of independent reactions is given by $R = C - M$. For systems having more than three components and two elements, it becomes very difficult to determine the independent reactions that define the system. BALANCE helps you determine the independent reactions quickly and easily.

To use BALANCE, you must first arrange the components according to their elemental abundances. As an example, consider the system (CH_4, H_2O, H_2, CO, CO_2) (CHO) shown in Table 3.1.

The data for the program are written as data statements starting at line 600 as shown below:

600 DATA CH4,1,4,0

610 DATA H2O,0,2,1

620 DATA H2,0,2,0

630 DATA CO,1,0,1

640 DATA CO2,1,0,2

After all the data are entered, type **RUN**. The program will prompt you for the number of components and the number of elements. Answer the

---

Smith, W; Missen, R., *Chemical Reaction Equilibrium Analysis,* John Wiley & Sons, New York, NY., 1982, pp. 18–26, 244–245.

Table 3.1. Coefficient Matrix

COMPOUND	C	H	O
CH_4	1	4	0
H_2O	0	2	1
H_2	0	2	0
CO	1	0	1
CO_2	1	0	2

prompts with **5,3**; the program then begins the calculation. The coefficient matrix is printed followed by the independent reactions. The first component printed is assumed to have a reaction coefficient of -1; that is, a reactant. The following components may have a positive, negative, or zero coefficient. A negative coefficient means that the component is a reactant. A positive coefficient means that the component is a product. A zero coefficient means that the component does not enter into the reaction.

EXAMPLE

READY

RUN

NO. OF COMPONENTS, NO. OF ELEMENTS? **5,3**

COEFFICIENT MATRIX

CH4	1	4	0
H2O	0	2	1
H2	0	2	0
CO	1	0	1
CO2	1	0	2

INDEPENDENT REACTIONS

CO	1 CH4	1 H20	-3 H2
CO2	1 CH4	2 H20	-4 H2

READY

LISTING FOR BASIC PROGRAM "BALANCE"

```
10 DIM X$(15),C(15,30)

20 CLS:INPUT"NO. OF COMPONENTS, NO. OF ELEMENTS";NS,M

30 LPRINT"COEFFICIENT MATRIX":LPRINT

40 FOR I=1 TO NS

50 READX$(I)

60 LPRINT X$(I);TAB(15);

70 FOR J=1 TO M

80 READ C(J,I)

90 LPRINT C(J,I);"      ";

100 NEXT J

110 LPRINT

120 NEXT I

130 LPRINT:LPRINT:LPRINT

140 REM-BASIC TRANSLATION OF FORTRAN PROGRAM PRESENTED BY SMITH,MISSEN
    , "CHEMICAL REACTION EQUILIBRIUM ANALYSIS", WITH PERMISSION, JOH
    N WILEY AND SONS

150 M1=M+1

160 FOR I=1 TO M

170 IP1=IP1+1

180 FOR K=I TO NS

190 FOR L=I TO M

200 IF C(L,K)<>0 THEN 260

210 NEXT L

220 NEXT K

230 M=I-1

240 M1=I
```

```
250 GOTO 490

260 IF K=I THEN GOTO 350

270 TEMP$=X$(K)

280 X$(K)=X$(I)

290 X$(I)=TEMP$

300 FOR J=1 TO M

310 TEMP=C(J,K)

320 C(J,K)=C(J,I)

330 C(J,I)=TEMP

340 NEXT J

350 IF L=I THEN 390

360 FOR J=I TO NS

370 C(I,J)=C(I,J)+C(L,J)

380 NEXT J

390 IF IP1>NS THEN 480

400 FOR L=1 TO M

410 IF L=I OR C(L,I)=0 THEN 470

420 R=C(L,I)/C(I,I)

430 FOR J=IP1 TO NS

440 C(L,J)=C(L,J)-C(I,J)*R

450 IF ABS(C(L,J))<1E-6 THEN C(L,J)=0

460 NEXT J

470 NEXT L

480 NEXT I

490 FOR I=1 TO M

500 FOR J=M1 TO NS

510 C(I,J)=C(I,J)/C(I,I)
```

```
520 NEXT J

530 NEXT I

540 NR=NS-M

550 CLS

560 LPRINT "INDEPENDENT REACTIONS":LPRINT

570 FOR J=M1 TO NS

580 LPRINT X$(J);"      ";

590 FOR I=1 TO M

600 LPRINT C(I,J);X$(I);"      ";

610 NEXT I

620 LPRINT

630 NEXT J

640 END

650 DATA CH4,1,4,0

660 DATA H2O,0,2,1

670 DATA H2,0,2,0

680 DATA CO,1,0,1

690 DATA CO2,1,0,2
```

CHEMICAL EQUILIBRIUM

EQUIL is a unique program that can calculate the equilibrium composition of a system consisting of a gas phase and a pure liquid or solid phase. EQUIL uses a first-order gradient projection method to minimize the free energy of a system described by the independent reactions. Hence, EQUIL is a stoichiometric algorithm. EQUIL can account for nonideality in the gas phase by calculating fugacity coefficients from available van der Waals constants. The solid or liquid phases are assumed to be pure and not to mix.

To illustrate the usefulness and simplicity of the program, consider the following problem.

PROBLEM: Due to an instrumentation failure, the steam flow to a methane/steam reformer was not at the indicated set point. Hot spots have developed in the reformer tubing, and the reformer must be shut down. Investigate the possibility of carbon being deposited on the reformer catalyst if it is known from the instrument department that the steam-to-carbon ratio was approximately 1:1 instead of the normal 3:1. The reformer operates at 1700 °F and 300 psig.

SOLUTION: The possible compounds present at equilibrium are CH_4, CO_2, CO, H_2, H_2O, and Cs. The independent reactions from BALANCE are

$$H_2O + 2CO = .5CH_4 + 1.5CO_2$$
$$H_2 + CO = .5CH_4 + .5CO_2$$
$$Cs + CO_2 = 2CO$$

The free energy of formations and the van der Waals constants are shown in Table 3.2
All data for the program are entered as data statements starting at line 4000. First, the reactions are entered using the form:

4000 REM-CHEMICAL REACTIONS

4010 DATA -1,H2O,-2,CO,.5,CH4,1.5,CO2

4020 DATA -1,H2,-1,CO,.5,CH4,.5,CO2

4030 DATA -1,CS,-1,CO2,2,CO

4040 REM-INGREDIENT CARDS NEXT. NAME FIRST, FREE ENERGY

NEXT, INITIAL MOLES.

4050 DATA CH4,9887,1

Table 3.2.[a]

COMPOUND	FREE ENERGY	A	B
CH_4	9887	2.253	.04278
CO_2	-94681	3.592	.84267
CO	-52049	1.485	.03985
Cs	0	—	—
H_2O	-43371	5.464	.03049
H_2	0	0.244	.03707

SOURCE: Data from Janaf tables and *Langes Handbook of Chemistry.*
[a]Temperature = 1200 K; Pressure (atm) = 21.41.

4060 DATA CO2, − 94681,0

4070 DATA CO, − 52049,0

4080 DATA CS,0,0

4090 DATA H2O, − 52049,1

4100 DATA H2,0,0

4110 REM-ENTER INITIAL GUESSES FOR EACH IN THE SAME ORDER

AS THE INGREDIENT CARDS.

4120 DATA .5,.10,.40,0,.40,1.60

4130 REM-ENTER VAN DER WAALS CONSTANTS FOR EACH

COMPOUND IN THE SAME ORDER AS EACH GASEOUS

COMPOUND. ENTER "A" FIRST, THEN "B" NEXT.

4140 DATA 2.53,.04278,3.592,.04267,

1.485,.03985,5.464,.03049,

.2444,.03707

Note: Solid and liquid compounds are denoted in the reaction and ingredient cards by a trailing S; that is, SIO2S or CACO3S. Initial guesses must satisfy the atomic abundances.

The results of the computation after 902 iterations are shown below.[1] The negative mole number for carbon indicates that carbon would have been consumed under the conditions specified. Another way of looking at the results is that the steam-to-carbon ratio must be approximately 1:1.2 before carbon would be deposited on the catalyst.

Equations for EQUIL

$$\Delta G_i = \Delta G_{fi} + R1T \ln \frac{n_i}{n_T} + R1T \ln (\phi_i P) \text{ (Gases Only)}$$

$$\Delta G_i = \Delta G_{fi} \text{ (Solids Only)}$$

$$\Delta G_{Tj} = \sum_{J=1}^{R} \sum_{i=1}^{R} v(j,i) \Delta G_i$$

---

[1]At times, EQUIL will converge very slowly. Reseeding the computer with new damping factors (larger) may help. If this does not help, the program should be compiled and run as machine code.

$$\xi_j = - \Delta G_{T_j} \lambda$$

$$\eta_{i+1} = \eta_i + \sum_{J=1}^{R} v(j,i) \xi_j \omega_i$$

$$\phi_i = \exp[(B_i P)/(.08207 * T)]$$
$$Bi = b_i - a_i/(.08207 * T)$$

$$\eta_T = \sum_{i=1}^{N} \eta_i \text{ (summation over each gaseous species)}$$

Nomenclature for EQUIL

VARIABLE	DESCRIPTION
η_i	Moles comp i
η_T	Total moles gas
ΔG_{fi}	Free energy of formation of comp i
ΔG_i	Free energy change of comp i
ΔG_{Tj}	Free energy change for each independent reaction j = 1 to R
ξ_j	Extent of reaction for reaction j
λ	Step size (a small number; that is, 1E-6)
ω_i	Step size parameter for comp i
θ_i	Fugacity coefficient for comp i
B_i	First viral coefficient for comp i
a_i	van der Waals constant for comp i
b_i	van der Waals constant for comp i
Rl	Gas constant (1.98 cal/mole-K)
T	Absolute temperature(K)

EXAMPLE

READY

RUN

Smith, W., Missen, R., *Chemical Reaction Equilibrium Analysis,* John Wiley & Sons, New York, NY, 1982, pp. 139–141.

Naphtali, L., *J. Chem. Phys.,* 1959, *31,* 263

Naphtali, L., *Ind. Eng. Chem.,* 1961, *53,* 287

Stull, D., Prophet, H., *Janaf Thermochemical Tables,* 2nd Ed., National Bureau of Standards, Washington D.C., 1971

Dean, J., *Lange's Handbook of Chemistry,* 11th Ed., McGraw Hill, New York, NY, 1973, pp. 11–27.

CHEMICAL EQUILIBRIUM

NO. OF INDEPENDENT REACTIONS:?   3

CHEMICAL SPECIES IN REACTION 1 =?   4

CHEMICAL SPECIES IN REACTION 2 =?   4

CHEMICAL SPECIES IN REACTION 3 =?   3

DAMPING FACTOR FOR RX 1 =?   1E-6

DAMPING FACTOR FOR RX 2 =?   1E-6

DAMPING FACTOR FOR RX 3 =?   1E-6

MAX. NO. OF ITERATIONS:?   1000

TOTAL NO. OF CHEMICAL SPECIES =?   6

SYSTEM TEMP(K), PRESSURE(ATM):?   1200,21.41

RELATIVE ERROR FOR SYSTEM(CAL/MOLE)=?   100

DO YOU HAVE VOLUMETRIC DATA(Y/N)?   Y

SOLUTION TO SYSTEM

COMPOUND	MOLES	RESIDUALS	INLET COMP
CH4	.3855	9.60625E-05	1
CO2	.0549544	2.8814E-04	0
CO	.651133	−3.84156E-04	0
CS	−.0915881	−4.72107E-08	0
H2O	.238958	−1.92125E-04	1
H2	1.99004	−5.1111E-12	0
MOLES GAS	3.32059		

G 1 = −96.0625   G 2 = −2.67969   G 3 = −96.6875

TEMP(K) = 1200   PRESS(ATM) = 21.41

NO. OF ITERATIONS = 902

LAMBDA(1) = 2.5E-07   LAMBDA(2) = 2.38419E-13

LAMBDA(3) = 9.76563E-10

CHEMICAL DATA		VAN DER WAALS DATA	
COMPOUND	FREE ENERGY	A	B
CH4	9887	2.253	.04278
CO2	−94681	3.592	.04267
CO	−52049	1.485	.03985
CS	0	0	0
H2O	−43371	5.464	.03049
H2	0	.2444	.03707

FUGACITY COEFFICIENTS

COMPOUND	PHI
CH4	1.00434
CO2	1.00135
CO	1.0054
H2O	.994582
H2	1.00755

READY

```
   LISTING FOR BASIC PROGRAM "EQUIL"

 5 DEFINT I,J,C

10 CLEAR 500:R1=1.98

20 DIM V$(10,10),V(10,10),A$(10)

30 CLS:PRINT"CHEMICAL EQUILIBRIUM"

40 INPUT"NO. OF INDEPENDENT REACTIONS:";R

50 FOR J=1 TO R

60 PRINT "CHEMICAL SPECIES IN REACTION";J;"=";

70 INPUT C(J)

80 NEXT J

90 CLS
```

```
100 FOR I=1 TO R:PRINT"DAMPING FACTOR FOR RX";I;"=";:INPUT LAM(I):NEXT
    I

110 INPUT"MAX. NO. OF ITERATIONS:";MAX

120 INPUT"TOTAL NO. OF CHEMICAL SPECIES=";N

122 INPUT"SYSTEM TEMP(K), PRESSURE(ATM):";T,P

124 INPUT"RELATIVE ERROR FOR SYSTEM(CAL/MOLE)=";E1

127 RT=R1*T

130 CLS:NT=0

140 FOR J=1 TO R

150 FOR I=1 TO C(J)

160 READ V(J,I),V$(J,I)

170 NEXT I

180 NEXT J

190 FOR I=1 TO N

200 READ A$(I),GF(I),N0(I)

220 NEXT I

222 FOR I=1 TO N

224 READ N(I)

226 IF RIGHT$(A$(I),1)="S" THEN 229

228 NT=NT+N(I)

229 NEXT I

230 INPUT"DO YOU HAVE VOLUMETRIC DATA(Y/N)";A$

232 IF A$="N" THEN 236

234 GOSUB 3000:GOTO 238

236 FOR I=1 TO N:PHI(I)=1:NEXT I

238 FOR J=1 TO R

240 GT(J)=0

250 FOR I1=1 TO C(J)
```

```
260 GOSUB 2000

270 NEXT I1

280 NEXT J

290 FOR J=1 TO R

295 PRINT GT(J),;

300 E(J)=-GT(J)*LAM(J)

310 NEXT J

312 PRINT

314 IF M=0 THEN 318

316 GOSUB 2600

318 GOSUB 2700

320 FOR I=1 TO N

325 X(I)=0

330 FOR J=1 TO R

340 FOR I1=1 TO C(J)

350 IF A$(I)=V$(J,I1) THEN X(I)=X(I)+V(J,I1)*E(J)

360 NEXT I1

370 NEXT J

380 NEXT I

385 W=1

390 FOR I=1 TO N

395 DX=N(I)+X(I)

398 IF RIGHT$(A$(I),1)="S" THEN 420

400 IF DX>0 THEN 420

410 W1=.99*N(I)/(N(I)-X(I))

415 IF W1<W THEN W=W1

420 NEXT I

430 NT=0
```

```
440 FOR I=1 TO N

450 N(I)=N(I)+W*X(I)

452 IF RIGHT$(A$(I),1)="S" THEN 460

454 NT=NT+N(I)

460 NEXT I

470 FLAG=0

480 FOR I=1 TO R

490 IF ABS(GT(I)))E1 THEN FLAG=1

500 NEXT I

510 IF FLAG=0 THEN 570

520 M=M+1

530 IF M)MAX THEN 550

540 GOTO 238

550 CLS

560 PRINT"NO CONVERGENCE":GOTO 575

570 CLS:PRINT "SOLUTION TO SYSTEM"

575 PRINT"COMPOUND","MOLES","RESIDUALS","INLET COMP."

580 FOR I=1 TO N

590 PRINT A$(I),N(I),X(I),N0(I)

600 NEXT I

605 PRINT

610 PRINT"MOLES GAS",NT

620 FOR I=1 TO R

630 PRINT"G";I;"=";GT(I),;

640 NEXT I

645 PRINT

650 INPUT X$

655 CLS
```

```
660 PRINT"TEMP(K)=";T,"PRESS(ATM)=";P

670 PRINT"NO. OF ITERATIONS=";M

680 FOR I=1 TO R:PRINT"LAMBDA(";I;")=";LAM(I),,:NEXT I

690 PRINT

700 PRINT "CHEMICAL DATA",,"VAN DER WAALS DATA"

710 PRINT"COMPOUND","FREE ENERGY","A","B"

720 FOR I=1 TO N

730 PRINT A$(I),GF(I),A(I),B(I)

740 NEXT I

750 INPUT A$

760 CLS

770 PRINT"FUGACITY COEFFICENTS"

780 PRINT"COMPOUND","PHI"

790 FORI=1 TO N

800 IF RIGHT$(A$(I),1)="S" THEN 820

810 PRINT A$(I),PHI(I)

820 NEXT I

830 END

2000 REM-CALCULATE FREE ENERGY CHANGES

2010 FOR I=1 TO N

2020 IF A$(I)<>V$(J,I1) THEN 2040

2022 IF RIGHT$(A$(I),1)="S" THEN 2035

2030 GT(J)=GT(J)+V(J,I1)*(GF(I)+RT*LOG(N(I)/NT)+RT*LOG(P)+RT*LOG(PHI(I)
     ))

2032 GOTO 2040

2035 GOSUB 2500

2040 NEXT I

2050 RETURN
```

```
2500 REM-CALC. ACTIVITY OF SOLIDS

2510 GT(J)=GT(J)+V(J,I1)*GF(I)

2520 RETURN

2600 REM-ADJUST STEP SIZE FOR EACH REACTION

2610 FOR I=1 TO R

2620 X=ABS((GP(I)-GT(I))/GP(I))

2630 IF X).10 THEN LAM(I)=LAM(I)/8:GOTO 2645

2640 LAM(I)=2*LAM(I)

2645 IF LAM(I)=0 THEN LAM(I)=1E-10

2650 NEXT I

2660 RETURN

2700 REM-SAVE PREVIOUS ESTIMATE OF GT(J)

2710 FOR I=1 TO R

2720 GP(I)=GT(I)

2730 NEXT I

2740 RETURN

3000 REM-COMPUTE FUGACITY COEFFICIENT

3010 FOR I=1 TO N

3020 IF RIGHT$(A$(I),1)="S" THEN 3060

3030 READ A(I),B(I)

3040 B1=B(I)-A(I)/(.08207*T)

3050 PHI(I)=EXP(B1*P/(.08207*T))

3060 NEXT I

3070 RETURN

4000 REM-CHEMICAL REACTION

4010 DATA -1,H2O,-2,CO,.5,CH4,1.5,CO2

4020 DATA -1,H2,-1,CO,.5,CH4,.5,CO2

4030 DATA -1,CS,-1,CO2,2,CO
```

```
4040 REM-INGREDIENT CARDS NEXT. NAME FIRST, FREE ENERGY NEXT, INITIAL M
     OLES

4050 DATA CH4,9887,1

4060 DATA CO2,-94681,0

4070 DATA CO,-52049,0

4080 DATA CS,0,0

4090 DATA H2O,-43371,1

4100 DATA H2,0,0

4110 REM-ENTER INITIAL GUESSES FOR EACH IN THE SAME ORDER AS THE INGRED
     IENT CARDS.

4120 DATA .5,.10,.40,0,.40,1.60

4130 REM-ENTER VAN DER WAALS CONSTANTS FOR EACH COMPOUND IN THE SAME OR
     DER AS EACH GASEOUS COMPOUND. ENTER "A" FIRST, THEN "B" NEXT.

4140 DATA 2.253,.04278,3.592,.04267,1.485,.03985,5.464,.03049,.2444,.03
     707
```

VAPOR LIQUID EQUILIBRIUM

The design of distillation towers requires accurate vapor liquid equilibrium (VLE). Often data are unavailable from the literature and there is not enough time to run a pilot trial before the cost of a new distillation tower can be estimated. The process engineer usually uses Raoult's law and hopes that deviations are small.

Solubility parameters can provide accurate first approximations for the VLE of nonpolar to slightly polar binary systems that are judged to show positive deviations from Raoult's law. VLE is a program designed to compute binary VLE using solubility parameters.

The program requires that you supply the constants to Antoine vapor pressure equations of the form

$$\log P = [-.2185 * A1/T(K) + B1]$$
$$\log P = \log(\text{base 10})$$
$$P = mmHg$$

Constants are available from the *CRC Handbook or Chemistry and Physics* or they may be calculated if you know the heat of vaporization (A1 = cal/mole) and the vapor pressure at one temperature. You must also supply the following information.

1. Name of each component
2. Molecular weights
3. Density of each component (gram/ml)
4. Heats of vaporization (cal/mole) or solubility parameter at 25 (C)
5. Pressure (atm)
6. Estimates of the boiling points of each pure component—T(°C)

The program calculates and prints out activity coefficients, boiling points, and VLE composition.

Equations for VLE

$$RT \ln (\gamma_1) = V_1 \theta_2^2 (\delta_1 - \delta_2)^2$$
$$RT \ln (\gamma_2) = V_2 \theta_1^2 (\delta_1 - \delta_2)^2$$
$$P_1^* = \exp[2.303(-.2185A_1/T + B1)]/760$$
$$P_2^* = \exp[2.303(-.2185A_2/T + B2)]/760$$
$$\theta_1 = V_1 X_1 / (V_1 X_1 + V_2 X_2)$$
$$\theta_2 = V_2 X_2 / (V_1 X_1 + V_2 X_2)$$
$$Y_1 = \gamma_1 X_1 P_1^*$$
$$Y_2 = \gamma_2 X_2 P_2^* \quad \text{or} \quad 1 - Y_1$$
$$\delta_1 = \sqrt{\delta^2 + \delta_H^2 + \delta_p^2} \quad \text{or} \quad \delta_1 = ((Hv_1 + RT)/V_1)^{1/2}$$
$$\delta_2 = \sqrt{\delta^2 + \delta_H^2 + \delta_p^2} \quad \text{or} \quad \delta_2 = [(Hv_2 + RT)/V_1]^{1/2}$$

Nomenclature for VLE

VARIABLE	DESCRIPTION
T	Temperature (K)
P_1^*	Vapor press comp 1 (atm)
P_2^*	Vapor press comp 2 (atm)
X_1	Liquid mole fraction comp 1
X_2	Liquid mole fraction comp 2
Y_1	Vapor mole fraction comp 1
Y_2	Vapor mole fraction comp 2
V_1	Molar volume comp 1 (ml/mole) at 25°C
V_2	Molar volume comp 2 (ml/mole) at 25°C
θ_1	Liq volume fraction comp 1
θ_2	Liq volume fraction comp 2
δ_1	Solubility parameter comp 1 at 25°C

Prausnitz, J., *Molecular Thermodynamics of Fluid Phase Equilibria,* Prentice-Hall, Englewood Cliffs, N.J., 1969, pp. 270–278.

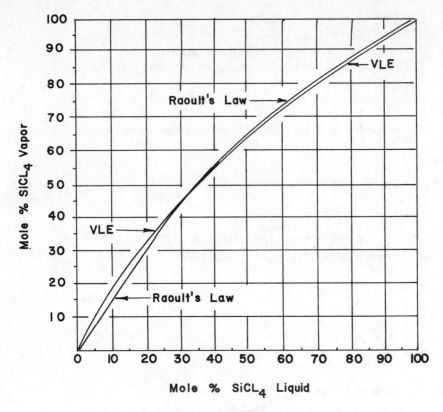

Figure 3.1 SiCl₄/hexane vapor-liquid equilibrium.

δ_2	Solubility parameter comp 2 at 25°C
MW	Mole weight
γ_1	Activity coefficient comp 1
γ_1	Activity coefficient comp 2
R	Gas constant (1.98 cal/mole-K)
HV_1, HV_2	Heat of vaporization (cal/mole)

EXAMPLE

READY

> RUN

BINARY VAPOR LIQUID EQUILIBRIUM

NAME OF COMPONENT NO. 1 = ? **HEXANE**

MOLE WT. = ?   **84**

NAME OF COMPONENT NO. 2 = ?   **SICL4**

MOLE WT. = ?   **170**

SYSTEM PRESSURE(ATM) = ?   **1**

COMPONENT NO.1 = HEXANE

ANTOINE CONSTANTS A1 = ?   **7627.2**

ANTOINE CONSTANT B1 = ?   **7.71712**

DO YOU HAVE SOLUBILITY PARAMETER(Y/N)?   **Y**

SOLUBILITY PARAMETER = ?   **7.3**

LIQUID DENSITY (GRAM/ML) = ?   **.659**

ESTIMATED BP AT SYSTEM PRESSURE = ?   **75**

COMPONENT NO.2 = SICL4

ANTOINE CONSTANTS A2 = ?   **7581.6**

ANTOINE CONSTANTS B2 = ?   **7.93122**

DO YOU HAVE SOLUBILITY PARAMETER(Y/N)?   **N**

HEAT OF VAPORIZATION(CAL/MOLE) = ?   **7581.6**

LIQUID DENSITY (GRAM/ML) = ?   **1.457**

ESTIMATED BP AT SYSTEM PRESSURE = ?   **56**

MAKE SURE LINE PRINTER IS SET UP

PRESS <ENTER> TO CONTINUE?

HEXANE	SICL4	PRESSURE (ATM) = 1
A1 = 7627.2	A2 = 7581.6	
B1 = 7.71712	B2 = 7.93122	

Hv (cal/mole) = 7627.2	7581.6	
MOLAR VOL. (ML/mole) = 127.466	116.678	
DENSITY(GRAM/ML) = .659	1.457	
SOL. PARAMETER = 7.3	8.36874	

X1	PHI	Y1	X2	PHI	Y2	T(C)
0	1.25119	0	1	1	1	54.9755
.02	1.23915	.0141961	.98	1.0001	.985804	55.1393
.04	1.22741	.0282914	.96	1.00039	.971712	55.3067
.06	1.21608	.0423019	.94	1.00087	.957698	55.4776
.08	1.20516	.0562442	.92	1.00154	.943758	55.6522
.1	1.19463	.0701338	.9	1.0024	.929867	55.8306
.12	1.18449	.0839875	.88	1.00345	.916024	56.013
.14	1.17471	.0978185	.86	1.00467	.902184	56.1987
.16	1.1653	.111645	.84	1.00608	.888356	56.3887
.18	1.15623	.125481	.82	1.00767	.874518	56.5827
.2	1.1475	.139343	.8	1.00944	.860655	56.7811
.22	1.1391	.153248	.72	1.01138	.846752	56.9838
.24	1.13103	.16721	.76	1.0135	.832791	57.191
.26	1.12326	.181245	.74	1.0158	.818753	57.4028
.28	1.1158	.195373	.72	1.01827	.804627	57.6195
.3	1.10863	.209609	.7	1.02091	.790398	57.8413
.32	1.10176	.223968	.68	1.02372	.776034	58.068
.34	1.09516	.238471	.66	1.0267	.761529	58.3002
.36	1.08884	.253136	.64	1.02986	.746865	58.538
.38	1.08278	.267981	.62	1.03318	.732018	58.7817
.4	1.07698	.283028	.6	1.03666	.716972	59.0315
.42	1.07144	.298296	.58	1.04032	.701705	59.2877

X1	PHI	Y1	X2	PHI	Y2	T(C)
.44	1.06614	.313806	.56	1.04414	.686192	59.5504
.46	1.06109	.329584	.54	1.04813	.670417	59.8201
.48	1.05627	.34565	.52	1.05228	.65435	60.097
.5	1.05169	.362031	.5	1.0566	.637969	60.3815
.52	1.04733	.378753	.48	1.06108	.621246	60.6739
.54	1.04319	.395845	.46	1.06572	.604156	60.9746
.56	1.03927	.413336	.44	1.07053	.586665	61.2841
.58	1.03555	.431257	.42	1.0755	.568743	61.6026
.6	1.03205	.449642	.4	1.08063	.550358	61.9307
.62	1.02875	.468527	.38	1.08592	.53147	62.2688
.64	1.02565	.487954	.36	1.09138	.512046	62.6175
.66	1.02274	.50796	.34	1.09699	.49204	62.9774
.68	1.02003	.528591	.32	1.10276	.47141	63.3489
.7	1.0175	.549894	.3	1.1087	.450105	63.7327
.72	1.01516	.571924	.28	1.11479	.428076	64.1294
.74	1.01299	.594734	.26	1.12104	.405265	64.5398
.76	1.01101	.618388	.24	1.12744	.381612	64.9647
.78	1.0092	.64295	.22	1.134	.35705	65.4048
.8	1.00756	.668494	.2	1.14072	.331506	65.8609
.82	1.00609	.695098	.18	1.14759	.304901	66.3341
.84	1.00479	.72285	.16	1.15461	.27715	66.8253
.86	1.00364	.751853	.14	1.16179	.248158	67.3359

X1	PHI	Y1	X2	PHI	Y2	T(C)

.88	1.00266	.782185	.12	1.16911	.217813	67.866
.9	1.00184	.81399	.1	1.17658	.186007	68.4181
.92	1.00117	.847391	.08	1.1842	.152609	68.9931
.94	1.00066	.882523	.06	1.19197	.117478	69.5924
.96	1.00029	.919547	.04	1.19987	.0804538	70.2177

| .98 | 1.00007 | .958642 | .02 | 1.20792 | .0413605 | 70.8707 |
| 1 | 1 | 1 | 0 | 1.2161 | 0 | 71.5531 |

DO YOU WANT TO CHANGE THE PRESSURE (Y/N)?   N

PROGRAM TERMINATED

READY

>

    LISTING FOR BASIC PROGRAM "VLE"

```
 10 CLEAR 1000

 20 CLS

 30 PRINT"BINARY VAPOR LIQUID EQUILIBRIUM"

 40 INPUT"NAME OF COMPONENT NO.1=";A1$

 50 INPUT"MOLE WT.=";M1

 60 INPUT"NAME OF COMPONENT NO. 2=";A2$

 70 INPUT"MOLE WT.=";M2

 80 INPUT"SYSTEM PRESSURE(ATM)=";P

 90 CLS

100 PRINT"COMPONENT NO.1=";A1$
```

```
110 INPUT"ANTOINE CONSTANTS A1=";A1

120 INPUT"ANTOINE CONSTANT B1=";B1

130 INPUT"DO YOU HAVE SOLUBILITY PARAMETER(Y/N)";Z1$

140 IF Z1$="N" THEN 160

150 INPUT"SOLUBILITY PARAMETER=";C1:GOTO 170

160 INPUT"HEAT OF VAPORIZATION(CAL/MOLE)=";H1

170 INPUT"LIQUID DENSITY(GRAM/ML)=";D1

180 INPUT"ESTIMATED BP AT SYSTEM PRESSURE=";T1

190 CLS

200 PRINT"COMPONENT NO.2=";A2$

210 INPUT"ANTOINE CONSTANTS A2=";A2

220 INPUT "ANTOINE CONSTTANTS B2=";B2

230 INPUT"DO YOU HAVE SOLUBILITY PARAMETER(Y/N)";Z2$

240 IF Z2$="N" THEN 260

250 INPUT"SOLUBILITY PARAMETER=";C2:GOTO 270

260 INPUT"HEAT OF VAPORIZATION(CAL/MOLE)=";H2

270 INPUT"LIQUID DENSITY (GRAM/ML)=";D2

280 INPUT"ESTIMATED BP AT SYSTEM PRESSURE=";T2

290 CLS

300 PRINT"MAKE SURE LINE PRINTER IS SET UP"

310 INPUT"PRESS <ENTER> TO CONTINUE";Z$

320 V1=M1/D1

330 IF Z1$="N"THEN GOSUB 880

340 V2=M2/D2

350 IF Z2$="N" THEN GOSUB 910

360 N=0

370 LPRINT A1$;TAB(40);A2$;TAB(60);"PRESSURE(ATM)=";P
```

```
380 LPRINT"A1=";A1;TAB(40);"A2=";A2

390 LPRINT"B1=";B1;TAB(40);"B2=";B2

400 LPRINT"------------------------------------------------------------
    ----------------"

410 LPRINT"HV (CAL/MOLE)=";A1;TAB(40);A2

420 LPRINT"MOLAR VOL.(ML/MOLE)=";V1;TAB(40);V2

430 LPRINT"DENSITY(GRAM/ML)=";D1;TAB(40);D2

440 LPRINT"SOL. PARAMETER=";C1;TAB(40);C2

450 LPRINT"------------------------------------------------------------
    ----------------"

460 LPRINT"X1";TAB(10);"PHI";TAB(20);"Y1";TAB(40);"X2";TAB(50);"PHI";T
    AB(60);"Y2";TAB(70);"T(C)"

470 LPRINT"------------------------------------------------------------
    ----------------"

475 X2=1:T=TZ:AB=1:GOSUB 670

480 FOR X=0 TO 100 STEP 2

490 N=N+1

500 X1=X/100:X2=1-X1

510 Z1=X1*V1/(X1*V1+X2*V2)

520 Z2=X2*V2/(X1*V1+X2*V2)

530 AA=EXP(Z2[2*V1*(C1-C2)[2/(1.98*(T+273.15)))

540 AB=EXP(Z1[2*V2*(C1-C2)[2/(1.98*(T+273.15)))

570 GOSUB 670

580 GOSUB 800

590 IF N )=5 THEN GOSUB 850

600 LPRINT X1;TAB(10);AA;TAB(20);Y1;TAB(40);X2;TAB(50);AB;TAB(60);Y2;T
    AB(70);T

610 NEXT X

620 INPUT"DO YOU WANT TO CHANGE THE PRESSURE(Y/N)";Z$

630 IF Z$="N" THEN 650
```

```
640 INPUT"NEW SYSTEM PRESSURE(ATM)=";P:GOTO 290

650 PRINT"PROGRAM TERMINATED"

660 END

670 TEMP=T

680 T=T+.0001

690 GOSUB 780

700 F0=F1

710 T=T-.0001

720 GOSUB 780

730 DERV=(F0-F1)/.0001

740 T=T-F1/DERV

750 IF ABS(TEMP-T)=<.001 THEN RETURN

760 TEMP=T

770 GOTO 680

780 F1=AA*X1/760*EXP((-.2185*A1/(T+273)+B1)*2.303)+AB*X2/760*EXP((-.21
    85*A2/(T+273)+B2)*2.303)-P

790 RETURN

800 P1=EXP((-.2185*A1/(T+273)+B1)*2.303)/760

810 P2=EXP((-.2185*A2/(T+273)+B2)*2.303)/760

820 Y1=AA*X1*P1/P

830 Y2=AB*X2*P2/P

840 RETURN

850 LPRINT"-----------------------------------------------------------
    --------------------"

860 N=0

870 RETURN

880 REM-CALCULATE SOL. PARM.

890 C1=SQR((H1+1.98*(298))/V1)
```

```
900 RETURN

910 REM-CALCULATE SOL. PARM.

920 C2=SQR((H2+1.98*(298))/V2)

930 RETURN
```

GAS SOLUBILITY

There is a shortage of good gas/liquid solubility data in the chemical literature. Common gases such as CO, N_2, CO_2, and CH_4 are frequently encountered gases in chemical processing for which very little data exist.

GASSOL is a computer program that will predict the solubility of 11 common gases in a pure liquid or a liquid mixture. The program uses solubility parameters to make the estimate. The 11 gases in the program are listed in Table 3.3. To use the program, you must supply the following information:

1. Name of the liquid components
2. Molecular weights of the liquid components
3. Density of the liquid components (gram/ml) at 25°C
4. Moles of each liquid in the liquid phase
5. Heats of vaporization (cal/mole) or solubility parameters at 25°C
6. Pressure (atm)
7. Temperature (°C)

The program allows you to select the desired gas from a menu and to change the temperature and pressure of the system. Appendix A.2 contains

Table 3.3. Gaseous Solutes

NUMBER	GAS SOLUTE
1	Nitrogen
2	Carbon monoxide
3	Oxygen
4	Argon
5	Methane
6	Carbon dioxide
7	Krypton
8	Ethylene
9	Ethane
10	Radon
11	Chlorine

a list of chemical compounds and their group solubility parameters. The solubility parameter needed by the program is the square root of the sum of the dispersal, polarization, and hydrogen-bonding parameters squared.

Equations for GASSOL

$$\frac{1}{X_2} = \frac{FL2}{P} * \exp\left[V_2^L(\delta_1 - \delta_2)^2\theta_1^2/RT\right]$$

$$FL2 = (9.79344 - 31.0059Tr + 32.1184Tr^2 - 11.6937Tr^3 + 1.43181Tr^4)\,[\exp(V_2^L(P - 1)/R1T)]*Pc$$

$$\delta_i = (\delta_D^2 + \delta_p^2 + \delta_H^2)^{1/2} \quad \text{or} \quad \delta_i = [(Hv + RT)/V_i^L]^{1/2}$$

$$\bar{\delta} = \sum_{i=1}^{n} \theta_i\delta_i \text{ (for liquid mixtures)}$$

$$\theta_1 = V_1^L X_1/(V_1^L X_1 + V_2^L X_2)$$

$$\theta_2 = V_2^L X_2/(V_1^L X_1 + V_2^L X_2)$$

$$V_i^L = MW/\rho_i$$

$$T_r = T/T_c$$

Nomenclature for GASSOL

VARIABLE	DESCRIPTION
Hv	Heat or vaporization (cal/mole)
T	Temperature (K)
T_r	Reduced temperature
T_c	Critical Temp (K)
P_c	Critical Pressure (atm)
P	Pressure (atm)
X_1	Mole fraction comp 1
X_2	Mole fraction comp 2 (gas)
V_1^L	Molar volume comp 1, ml/mole
V_2^L	Molar volume comp 2, ml/mole (gas)
δ_1	Solubility parameter comp 1 (cal/ml)$^{1/2}$
δ_2	Solubility parameter comp 2 (cal/ml)$^{1/2}$ (gas)
$\bar{\delta}$	Solubility parameter mix (cal/ml)$^{1/2}$
θ_1	Volume fraction comp 1
θ_2	Volume fraction comp 2 (gas)
MW	Mole weight
P	Pressure (atm)
ρ_1	Liquid density comp 1 (gram/ml)

Prausnitz, J., *Molecular Thermodynamics of Fluid Phase Equilibria,* Prentice-Hall, Englewood Cliffs, N.J., 1969, pp. 365–372.

VARIABLE	*DESCRIPTION*
FL2	Hypothetical liquid fugacity (atm) (gas)
R	Gas constant (1.98 cal/mole -K)
R1	Gas constant (82.06 ml/mole-K-atm)
H_v	Heat of vaporization of liquid solvent (cal/mole)

EXAMPLE

RUN "GASSOL"

ESTIMATION OF GAS SOLUBILITY

NO.	GASEOUS SOLUTE
1	NITROGEN
2	CARBON MONOXIDE
3	OXYGEN
4	ARGON
5	METHANE
6	CARBON DIOXIDE
7	KRYPTON
8	ETHYLENE
9	ETHANE
10	RADON
11	CHLORINE

CHOOSE SOLUTE? **5**

NO. OF PURE COMPONENTS IN THE LIQUID PHASE=? **1**

COMPONENT NO. 1

NAME=? **HEXANE**

MOLE WT.=? **84**

LIQUID DENSITY (GRAM/ML)=? **.638**

DO YOU HAVE SOLUBILITY PARAMETER(Y/N)? **Y**

SOLUBILITY PARAMETER = ?   **7.3**

SYSTEM PRESSURE (ATM) = ?   **1.5**

SYSTEM TEMP (C) = ?   **50**

****PROGRAM RUNNING****

MAKE SURE LINE PRINTER IS SET UP?

SYSTEM PRESSURE (ATM) = 1.5

SYSTEM TEMP (K) = 323.15

COMPOUND:HEXANE

MOLE% = .99397

MW = 84

DENSITY (GRAM/ML) = .638

MOLAR VOLUME = 131.661

SOLUBILITY PARAMETER = 7.3

HV-CAL/MOLE = 0

GASEOUS SOLUTE = METHANE

CRITICAL TEMP (K) = 190.65

CRITICAL PRESSURE (ATM) = 45.8

CALCULATED SOLUBILITY (MOLE%) = 6.03012E-03

ANOTHER RUN (Y/N)?   **N**

PROGRAM EXECUTION TERMINATED

READY


```
    LISTING FOR BASIC PROGRAM "GASSOL"

 10 R=1.98:R1=82.06

 20 DIM N(10),VL(11),TC(11),PC(11),G$(11),S(11)

 30 CLS
```

```
 40 FOR I=1 TO 11

 50 READ G$(I),VL(I),S(I),TC(I),PC(I)

 60 NEXT I

 70 PRINT"ESTIMATION OF GAS SOLUBILITY"

 80 PRINT"NO.","GASEOUS SOLUTE"

 90 FOR I=1 TO 11

100 PRINT I,G$(I)

110 NEXT I

120 INPUT"CHOOSE SOLUTE";Z

130 CLS:IF A$="Y" THEN T=T-273.15:GOTO 310

140 INPUT"NO. OF PURE COMPONENTS IN THE LIQUID PHASE=";N

150 FOR I=1 TO N

160 PRINT"COMPONENT NO. ";I

170 INPUT"NAME=";N$(I)

180 INPUT"MOLE WT.=";MW(I)

190 INPUT"LIQUID DENSITY (GRAM/ML)=";PL(I)

200 INPUT"DO YOU HAVE SOLUBILITY PARAMETER(Y/N)";A1$

210 IF A1$="N" THEN INPUT"HEAT OF VAPORIZATION(CAL/MOLE)=";HV(I)

220 VN(I)=MW(I)/PL(I)

230 U=(HV(I)+R*298)/VN(I)

240 IF A1$="Y" THEN 260

250 SP(I)=SQR(U)

260 INPUT"SOLUBILITY PARAMETER=";SP(I)

270 IF N=1 THEN 290

280 INPUT"NO. OF MOLES IN MIXTURE=";N(I)

290 CLS

300 NEXT I

310 INPUT"SYSTEM PRESSURE (ATM)=";P
```

```
320 INPUT"SYSTEM TEMP (C)=";T

330 T=T+273.15

340 CLS

350 PRINT"**** PROGRAM RUNNING ****"

360 GOSUB 720

370 CLS

380 INPUT"MAKE SURE LINE PRINTER IS SET UP";X$

390 LPRINT"SYSTEM PRESSURE (ATM)=";P

400 LPRINT"SYSTEM TEMP (K)=";T

410 LPRINT

420 FOR I=1 TO N

430 IF N=1 THEN X0=1-X2:GOTO 450

440 X0=N(I)/NT

450 LPRINT"COMPOUND=";N$(I)

460 LPRINT"MOLE%=";X0

470 LPRINT"MW=";MW(I)

480 LPRINT"DENSITY(GRAM/ML)=";PL(I)

490 LPRINT"MOLAR VOLUME=";VN(I)

500 LPRINT"SOLUBILITY PARAMETER=";SP(I)

510 LPRINT"HV-CAL/MOLE=";HV(I)

520 LPRINT:LPRINT

530 NEXT I

540 LPRINT:LPRINT

550 LPRINT"GASEOUS SOLUTE=";G$(Z)

560 LPRINT"CRITICAL TEMP (K)=";TC(Z)

570 LPRINT"CRITICAL PRESSURE (ATM)=";PC(Z)

580 IF N=1 THEN 610

590 LPRINT"CALCULATED SOLUBILITY (MOLE%)=";NG/NT
```

```
600 GOTO 620

610 LPRINT"CALCULATED SOLUBILITY (MOLE%)=";X2

620 LPRINT:LPRINT:LPRINT

630 INPUT"ANOTHER RUN (Y/N)";A$

640 IF A$="Y" THEN 670

650 PRINT"PROGRAM EXECUTION TERMINATED"

660 END

670 INPUT"TRY ANOTHER GAS (Y/N)";A$

680 IF A$="Y" THEN CLS:GOTO 80

690 INPUT"TRY ANOTHER TEMP OR PRESSURE (Y/N)";A$

700 IF A$="Y" THEN CLS:T=T-273.15:GOTO 310

710 GOTO 650

720 X=T/TC(Z)

730 Y=9.79344-31.0059*X+32.1184*X[2-11.6937*X[3+1.43181*X[4

740 X1=EXP(VL(Z)*(P-1)/(R1*T)):FL2=Y*X1*PC(Z)

750 IF N=1 THEN 960

760 TEMP=0:NT=0:NG=0

770 NT=0

780 FOR I=1 TO N

790 NT=NT+N(I)

800 NEXT I

810 NT=NT+NG:PHI=0:VXT=0

820 FOR I=1 TO N

830 VXT=VXT+N(I)/NT*VN(I)

840 NEXT I

850 VXT=VXT+NG*VL(Z)/NT

860 FOR I=1 TO N

870 V1=N(I)/NT*VN(I)/VXT
```

```
880 PHI=PHI+V1*SP(I)

890 NEXT I

900 PHI=PHI+NG/NT*VL(Z)*S(Z)/VXT

910 X=EXP(VL(Z)*(S(Z)-PHI)[Z/(R*T))*FL2/P

920 NG=NT/X

930 IF ABS(TEMP-NG)<=1E-7 THEN RETURN

940 TEMP=NG

950 GOTO 770

960 TEMP=0:XZ=0

970 X1=1-XZ

980 V1=X1*VN(1)/(X1*VN(1)+XZ*VL(Z))

990 X=EXP(VL(Z)*(S(Z)-SP(1))[Z*V1[Z/(R*T))*FL2/P

1000 XZ=1/X

1010 IF ABS(TEMP-XZ)<=1E-7 THEN RETURN

1020 TEMP=XZ

1030 GOTO 970

1040 DATA NITROGEN,32.4,2.58,126.05,33.5

1050 DATA CARBON MONOXIDE,32.1,3.13,134.15,35.0

1060 DATA OXYGEN,33.,4.0,154.35,49.7

1070 DATA ARGON, 57.1,5.33,151.15,48.0

1080 DATA METHANE,52,5.68,190.65,45.8

1090 DATA CARBON DIOXIDE,55,6.0,304.25,73.

1100 DATA KRYPTON,65,6.4,209.4,54.3

1110 DATA ETHYLENE,65,6.6,282.85,50.5

1120 DATA ETHANE,70,6.6,305.25,48.8

1130 DATA RADON,70,8.83,377.15,62.0

1140 DATA CHLORINE,74,8.70,417.15,76.1
```

4
MASS TRANSFER

MULTICOMPONENT DISTILLATION

DISTILL is a shortcut for the design of multicomponent distillation towers. The program is based on the FUEM method as set forth in the 5th edition of *Perry's Chemical Engineer's Handbook*. It requires you to supply antoine constants for the vapor pressure equations for each component. The equations are of the form

$$\log P = [-.2185 * A1/T(K) + B1]$$
$$\log P = \log P \text{ (base 10)}$$
$$P = mmHg$$

Constants are available from the *Handbook of Chemistry and Physics*, or they may be calculated if the vapor pressure at one point and the heat of vaporization are known. The constant A1 is the heat of vaporization (cal/mole) in the antoine equation. Data are entered in line 9000 in the following order:

9000 DATA NAME,A1,B1

Each component should have a separate data statement. After the data statements have been typed, the program will ask for the following information:

1. Number of components
2. Names of the light key (LK) and the heavy key (HK)
3. Feed composition (decimal mole%)
4. Estimated boiling points of the feed, distillate, and bottoms (°C)
5. Percentage recovery of the LK and HK

6. System pressure (ATM)
7. R/Rm; ie., 1.25 * Rm
8. Q value of feed

The program calculates the following parameters:

1. Distillate composition
2. Bottoms composition
3. Top and bottom temperature
4. Minimum stages, minimum reflux ratio
5. Actual stages
6. Feed plate location
7. Relative volatilites of feed, distillate, bottoms

As an example, consider the system of $SICL_4$, $SIHCL_3$ (TCS), CH_3SICL_3 (MTCS), and CH_4CL_2SI (MEH). Rectify a feed stream consisting of 10% (mole%) $SICL_4$, 10% TCS, 10% MTCS, and 70% MEH at 1 atm pressure. Recover 90% of the TCS (LK) in the over head and 99% of the $SICL_4$ (HK) in the bottoms. The computer's output follows the equations and nomenclature for DISTILL.

Equations for DISTILL

Fenske Equation[1]

$$Nm = \ln[(D_{LK}/D_{HK}) * (B_{HK}/B_{LK})]/\ln(K_{LK})$$

$$D_i = (D_{LK}/B_{HK} * z_i * K_i^{NM}) * [1 + (D_{LK}/B_{HK})K_i^{NM}]$$

$$B_i = Zz_i - D_i : y_i = P_i/D : x_i = B_i/B$$

$$P_D^* = Pt/ \sum_{i=1}^{N} y_iK_i \text{ Distillate dew point (T1)}$$

$$P_B^* = Pt/ \sum_{i=1}^{N} x_iK_i \text{ Bottom bubble point (T2)}$$

$$P_Z^* = Pt/ \sum_{i=1}^{N} z_iK_i \text{ Feed bubble point (T3)}$$

$$K_i = (K1_i * K2_i * K3_i)^{1/3}$$

Underwood Equations[1]

$$Rm + 1 = \sum_{i=1}^{N} K_ix_i/(K_i - \theta)$$

---

[1]Perry, R., Chilton, C., *Chemical Engineer's Handbook,* 5th Ed., McGraw-Hill, New York, NY, 1973, pp. *13–28*, 30.

$$1 - q = \sum_{i=1}^{N} K_i z_i / (K_i - \theta)$$

Gilliland Correlation[2]

$$(N - Nm)/(N + 1) = .75 - .75 \, [(R - Rm)/(R + 1)]^{.5668}$$

Vapor Pressure Equations

$$P_i^* = (-.2185 * Al_i/T + B1_i), \text{ where } T = K°$$

Feed Plate Location[3]

$$\left(\frac{M}{P} \right) = \left[\left(\frac{B}{D} \right) \left(\frac{Z_{HK}}{Z_{LK}} \right) \left(\frac{B_{LK}}{D_{HK}} \right)^2 \right]^{.206}$$

Nomenclature for DISTILL

VARIABLE	DESCRIPTION
Nm	Minimum stages at total reflux
D_{LK}	Moles of light key in distillate
D_{HK}	Moles of heavy key in distillate
B_{LK}	Moles of light key in bottoms
B_{HK}	Moles of heavy key in bottoms
Z	Moles/hr of feed
K_{LK}	Relative volatility of light key
Ki	Effective relative volatility of comp i
$K1_i$	Relative volatility at top of tower
$K2_i$	Relative volatility at feed plate
$K3_i$	Relative volatility at bottom of tower
P_D^*, T1	Dew point of distillate (K)
P_Z^*, T2	Bubble point of feed (K)
P_B^*, T3	Bubble point of bottoms (K)
P_i^*	Vapor pressure of comp i (atm)
Al_i	Antoine constant (cal/gram-mole)
$B1_i$	Antoine constant
R	Reflux Ratio
Rm	Minimum reflux ratio
N	Actual number of stages
θ	Root to Underwood equation

[2]Eduljee, H., Hydro. Proc., Sept., 1975, pp. 120–122.

[3]Ngai, I., *Chem. Eng.* Aug. 20, 1984, pp. 145–149, eq. (21)

VARIABLE	*DESCRIPTION*
q	Thermal condition of feed
D_i	Moles distillate of comp i
B_i	Moles bottoms of comp i
P	Total pressure (atm)
N_F	Feed plate location (0 = top)
y_i	Mole % comp i in vapor
x_i	Mole % comp i in liquid
z_i	Mole % comp i in feed
D	Total moles distillate
B	Total moles bottoms
M	Number of plates above feed plate
P	Number of plates below feed plate
Z	Total moles of feed

EXAMPLE

RUN

MULTICOMPONENT DISTILLATION

TYPE PROBLEM TITLE:?   **SILANE DISTILLATION**

TYPE NUMBER OF COMPONENTS?   **4**

FEED COMPOSITION (DECIMAL)

SICL4 = ?   **.10**

TCS = ?   **.10**

MTCS = ?   **.10**

MEH = ?   **.70**

TYPE THE NAME OF LIGHT KEY?   **TCS**

% RECOVERY LIGHT KEY-(DECIMAL) = ?   **.90**

TYPE NAME OF HEAVY KEY?   **SICL4**

% RECOVERY HEAVY KEY-(DECIMAL) = ?   **.99**

TYPE RATE * MINREFLUX = ?   **1.50**

TYPE Q VALUE FOR FEED = ?   **1**

TYPE MOLES PER HOUR FEED = ?   **100**

TYPE COLUMN PRESSURE (ATM) = ?   **1**

TYPE ESTIMATED TEMPS(C)

DISTILLATE = ?   **35**

FEED BUBBLE POINT = ?   **45**

BOTTOMS = ?   **55**

PROBLEM ID:SILANE DISTILLATION

FEED COMPOSITION

COMPONENT	MOLE%	RELATIVE VOLATILITY
SICL4	.1	1
TCS	.1	2.48586
MTCS	.1	.724721
MEH	.7	1.76574

% RECOVERY LIGHT KEY = 90   LK = TCS

% RECOVERY HEAVY KEY = 99   HK = SICL4

MIN. L/D = 1.29543   MIN. STAGES = 7.45902

R/RM = 1.5   ACTUAL L/D = 1.94315

REQ'D STAGES = 14.8926   FEED AT PLATE NO.: 11.02935 (0 = TOP)

	DISTILLATE		BOTTOMS	
COMPOUND	MOLES	XD	MOLES	XB
SICL4	.100001	2.63327E-03	9.9	.159615
TCS	9	.236993	1	.0161227
MTCS	9.14071E-03	2.40699E-04	9.99086	.16108
MEH	28.8666	.760133	41.1334	.663182

TOTALS   37.9758   62.0242

TOP TEMP = 37.3743   PRESSURE (ATM) = 1

FEED TEMP = 41.3279   Q VALUE = 1

BOTTOM TEMP = 44.1499

DO YOU WANT TO CHANGE COLUMN CONDITIONS(Y/N)?   N

PROGRAM EXECUTION TERMINATED

READY

```
     LISTING FOR BASIC PROGRAM "DISTILL"

 10 DIM A$(10),B(10),D(10),K(10),P(10),Z(10),X(10),Y(10),T(3),K1(10),K
    2(10),A(10),A1(10),K3(10)

 20 CLS

 30 PRINT"MULTICOMPONENT DISTILLATION"

 40 INPUT"TYPE   PROBLEM TITLE:";TITLE$

 50 CLS

 60 INPUT"TYPE   NUMBER OF COMPONENTS";N

 70 FOR I=1 TO N

 80 READ A$(I),A1(I),A(I)

 90 NEXT I

 92 IF Z1$="Y"THEN 94 ELSE 98

 94 INPUT"CHANGE FEED COMPOSTION (Y/N)";Z2$

 96 IF Z2$="N"THEN 160

 98 PRINT"FEED COMPOSTION (DECIMAL)"

100 FOR I=1 TO N

102 PRINT A$(I);"=";

103 INPUT Z(I)

104 NEXT I

106 CLS

160 INPUT"TYPE   THE NAME OF LIGHT KEY";B$

170 INPUT"% RECOVERY LIGHT KEY-(DECIMAL)=";RL

175 CLS

180 INPUT"TYPE   NAME OF HEAVY KEY";C$
```

```
190 INPUT"% RECOVERY HEAVY KEY-(DECIMAL)=";RH

195 CLS

200 INPUT"TYPE   RATE*MINREFLUX=";R1

210 INPUT"TYPE   Q VALUE FOR FEED=";Q

220 INPUT"TYPE   MOLES PER HOUR FEED=";Z

230 INPUT"TYPE   COLUMN PRESSURE(ATM)=";P

240 CLS

250 PRINT"TYPE   ESTIMATED TEMPS(C)"

260 INPUT"DISTILLATE=";T(1)

270 INPUT"FEED BUBBLE POINT=";T(2)

280 INPUT"BOTTOMS=";T(3)

290 CLS

300 FOR I=1 TO N

310 IF A$(I)=B$ THEN LK=I

320 IF A$(I)=C$ THEN HK=I

330 NEXT I

340 FOR I=1 TO 3

350 T=T(I)

360 GOSUB 3000   CALC VAPOR PRESSURES

370 ON I GOSUB 4000,5000,6000

380 NEXT I

390 FOR I=1 TO N

400 K(I)=(K1(I)*K2(I)*K3(I))[.333

410 NEXT I

440   D(HK)=Z*Z(HK)*(1-RH)

450 B(HK)=Z*Z(HK)*RH

460 D(LK)=Z*Z(LK)*RL
```

```
470 B(LK)=Z*Z(LK)*(1-RL)

480 NM=LOG((D(LK)/D(HK))/B(LK)*B(HK))/LOG(K(LK))

490 X=D(HK)/B(HK):D=0:B=0

500 FOR I=1 TO N

510 Y=K(I)[NM

520 D(I)=(X*Y*Z*Z(I))/(1+X*Y)

530 D=D+D(I)

540 B(I)=Z*Z(I)-D(I)

550 B=B+B(I)

560 NEXT I:X1=0:Y1=0:Z1=0

570 FOR I=1 TO N

580 Y(I)=D(I)/D

590 Y1=Y1+Y(I)*K(I)

600 X(I)=B(I)/B

610 X1=X1+X(I)*K(I)

620 Z1=Z1+Z(I)*K(I)

630 NEXT I

640 GOSUB 7000

650 IF ABS(T(3)-T3)=<.005 THEN 690

670 T(1)=T1:T(2)=T2:T(3)=T3

680 GOTO 340

690 X0=(K(LK)+1.0)/Z

700 C=1-Q:TEMP=X0

710 F1=0:F2=0

720 FOR I=1 TO N

730 F1=F1+(K(I)*Z(I))/(K(I)-X0)

740 NEXT I
```

```
745 F1=F1-C

750 X0=X0+.0001

760 FOR I=1 TO N

770 F2=F2+(K(I)*Z(I))/(K(I)-X0)

780 NEXT I

785 F2=F2-C

790 X0=X0-.0001

800 DERV=(F2-F1)/.0001

810 X0=X0-.10*F1/DERV

820 IF ABS((TEMP-X0)/X0)=(.001 THEN 850

830 TEMP=X0

840 GOTO  710

850 C1=0

860 FOR I=1 TO N

870 C1=C1+K(I)*Y(I)/(K(I)-X0)

880 NEXT I

890 RM=C1-1

900 R=RM*R1

910 GOSUB 8000

920 N1=((B/D)*(Z(HK)/Z(LK))*(B(LK)/D(HK)))[Z)[.206

930 N3=N2/(N1+1):NF=N2-N3

940 PRINT"PROBLEM ID:";TITLE$

950 PRINT

960 PRINT"FEED COMPOSITION"

970 PRINT"COMPONENT","MOLE%","RELATIVE VOLATILITY"

980 FOR I=1 TO N

990 PRINT A$(I),Z(I),K(I)

1000 NEXT I
```

```
1005 PRINT"% RECOVERY LIGHT KEY=";RL*100;"    LK=";A$(LK)

1007 PRINT"% RECOVERY HEAVY KEY=";RH*100;"    HK=";A$(HK)

1030 PRINT"MIN. L/D=";RM;TAB(30);"MIN. STAGES=";NM

1040 PRINT"R/RM=";R1;TAB(30);"ACTUAL L/D=";R

1050 PRINT"REQ'D STAGES=";NZ;TAB(30);"FEED AT PLATE NO.:";NF;"(0=TOP)"

1060 INPUT Z$

1065 CLS

1070 PRINT TAB(12);"DISTILLATE";TAB(42);"BOTTOMS"

1080 PRINT "COMPOUND";TAB(12);"MOLES";TAB(31);"XD";TAB(41);"MOLES";TAB(
     51);"XB"

1090 FOR I=1 TO N

1100 PRINT A$(I);TAB(12);D(I);TAB(27);Y(I);TAB(39);B(I);TAB(49);X(I)

1110 NEXT I

1115 PRINT"----------------------------------------------------------"

1120 PRINT"TOTALS";TAB(12);D;TAB(39);B

1130 IF N)5 THEN INPUT Z$:CLS

1135 PRINT

1140 PRINT"TOP TEMP="T1;TAB(30);"PRESSURE(ATM)=";P

1150 PRINT"FEED TEMP=";T2;TAB(30);"Q VALUE=";Q

1160 PRINT"BOTTOM TEMP=";T3

1165 INPUT Z$

1170 CLS

1180 INPUT"DO YOU WANT TO CHANGE COLUMN CONDITIONS(Y/N)";Z1$

1190 IF Z1$="Y" THEN CLS:GOTO 92

1200 PRINT"PROGRAM EXECUTION TERMINATED"

1210 END

3000 FOR I1=1 TO N

3010 P(I1)=EXP((-.2185*A1(I1)/(T+273)+A(I1))*2.303)/760
```

```
3020 NEXT I1

3030 RETURN

4000 FOR I1=1 TO N

4010 K1(I1)=P(I1)/P(HK)

4020 NEXT I1

4030 RETURN

5000 FOR I1=1 TO N

5010 K2(I1)=P(I1)/P(HK)

5020 NEXT I1

5030 RETURN

6000 FOR I1=1 TO N

6010 K3(I1)=P(I1)/P(HK)

6020 NEXT I1

6030 RETURN

7000 P2=P/Y1

7010 GOSUB 7900

7020 T1=T0

7030 P2=P/Z1

7040 GOSUB 7900

7050 T2=T0

7060 P2=P/X1

7070 GOSUB 7900

7080 T3=T0

7090 RETURN

7900 T0=(-.2185*A1(HK))/(LOG(P2*760)/2.303-A(HK))-273

7910 RETURN

8000 X2=.75-.75*((R-RM)/(R+1))[.5668
```

```
8010 NZ=(1+NM)/(1-XZ)

8020 RETURN

9000 DATA SICL4,7581.6,7.93122

9010 DATA TCS,6878.2,7.83744

9020 DATA MTCS,7450,7.69968

9030 DATA MEH,7011.1,7.78121
```

SIEVE PLATE EFFICIENCY

The *AIChE Bubble Tray Manual*[1] is probably the most authoritative work available to predict the mass transfer efficiency of bubble cap and sieve plates. TRYEFF is an adaptation[2] of the AIChE method that can be used for single pass sieve plates.[3]

The program requires that you supply the following information about the sieve plate:

1. Tray diameter (ft)
2. Weir length (ft)
3. Weir height (in.)
4. Active area (ft^2)
5. Liquid path length (weir to weir in ft)
6. Collapsed froth height (in.)
7. Liquid and gas rates (pph)
8. Fluid densities (lb/ft^3)
9. Gas and liquid viscosities (CP)

The program also prompts you to supply gas and liquid diffusivities. If you do not have the diffusion coefficients, the program will ask you to supply the following information, which is necessary to make an estimate:

1. Molecular weights
2. Atomic voluumes (from Table 4.1 A–C).

---

[1]Bubble Tray Design Manual, American Institute of Chemical Engineers, New York, NY, 1958.

[2]Treybal, R., Mass Transfer Operations, 2nd Ed., McGraw Hill, New York, NY, 1969, pp. 146–150.

[3]For Tray nomenclature see Fig. 4.1 and Fig. 4.2.

Table 4.1A.   Atomic Volumes[4]

GAS DIFFUSIVITY	
ELEMENT	VOLUME
C	16.50
H	1.98
O	5.48
N	5.69
CL	19.50
S	17.00
Aromatic ring	−20.20
Heterocyclic ring	−20.20

3. Viscosity of gas and liquid (cp)
4. Solvent association parameter (liquids)

Atomic volumes are shown in Table 4.1A–C. Solvent association parameters are listed in Table 4.2 The atomic volumes for each compound is obtained by summing the atomic volume of each element in the compound. For example, the atomic volume of propane (C_3H_8) is 3(16.50) + 8(1.98) = 63.9. Notice that there are two listings of atomic volumes. Tables 4.1A and 4.1B are to be used for calculating atomic volumes of compounds that diffuse as gases. Table 4.1C is to be used for diffusing solutes in liquids.

Table 4.1B.   Diffusional Volumes for Simple Molecules—
Gas Diffusivity[4]

GAS	+V	GAS	+V
H_2	7.07	CO	18.9
D_2	6.70	CO_2	26.9
HE	2.88	N_2O	35.9
N_2	17.9	NH_3	14.9
O_2	16.6	H_2O	12.7
AR	16.1	CCL_2F_2	114.8
KR	22.8	SF_6	69.7
XE	37.7	CL_2	37.7
BR_2	67.2	SO_2	41.1
Air	20.1		

Table 4.1C.[4]

LIQUID ATOMIC VOLUMES	
ELEMENT	VOLUME
C	14.8
H	3.7
CL	24.6
BR	27.0
I	37.0
N	15.6
S	25.6
N in primary amines	10.5
N in secondary amines	12.0
O	7.40
O in methyl esters	9.1
O in higher esters	11.0
O in acids	12.0
O in methyl ethers	9.9
O in higher ethers	11.0
Benzene ring-subtract	15.0
Naphthalene ring-subtract	30.0

Equations for TRYEFF

$$NG = (.776 + .116 * HW - .290 * VA * PG^{.5} + 9.72 * q/Z)/Sc^{.5}$$
$$NL = 7.31E5 * DL^{.5} * (.26 * VA * PG^{.5} + .15) * THETA$$
$$De = (.774 + 1.026 * VA + 67.2 * q/z + .90 * HW)^2$$
$$THETA = (2.31E-5 * HL * z * Z)/q$$
$$1/Nt_{og} = 1/NG + m * G/(L * NL)$$
$$E_{og} = 1 - exp(-Nt_{og})$$
$$n = Pe/z * [(1 + 4 m * G * E_{og}/(L * Pe))^{.5} - 1]$$
$$Pe = Z^2/(De * THETA)$$
$$X1 = n + Pe$$
$$X2 = 1 - exp(- (n + Pe))$$

Table 4.2. Solvent Association Parameters[4]

SOLVENT	PHI
Water	2.6
Methanol	1.9
Ethanol	1.5
Nonpolar solvents	1.0

[4]Perry, R., Chilton, C., *Chemical Engineer's Handbook,* 5th Ed., McGraw-Hill, New York, 1973, pp. 3–234, 3–229.

$X3 = \exp(n) - 1$

$Y1 = X1 * (1 + X1/n)$

$Y2 = n * (1 + n/X1)$

$E_{mg}/E_{og} = (X2/Y1 + X3/Y2)$

$E_{mge} = E_{mg}/\{1 + E_{mg} * [E/(1 - E)]\}$

$z = (D + W)/2$

$Sc = (u_g * 2.42)/(PG * DG)$

$VA = G/(3600 * PG * AA)$

$DG = A/B$

$A = .001 * T^{1.75} * (1/M1 + 1/M2)^2$

$B = P * (V1^{.333} + V2^{.333})^2$

$DL = 7.4E - 8 * (PHI * MW)^{.5} * T/(u_l * V^{.6})$

$q = L/(3600 * PL)$

Nomenclature for TRYEFF

VARIABLE	DESCRIPTION
NG	Number of gas transfer units
NL	Number of liquid transfer units
De	Eddy diffusivity (ft^2/hr)
THETA	Residence time on tray (sec)
Nt_{og}	Overall gas transfer units
E_{og}	Point gas phase efficiency
n	Calculated constant
Pe	Peclet number
E_{mg}	Murphee gas phase efficiency
E_{mge}	Murphee gas phase efficiency corrected for entrainment
Sc	Schmidt number—gas
DG	Gas diffusivity (cm^2/sec) * 3.88 = (ft^2/hr)
DL	Liquid diffusivity (cm^2/sec) * 3.88 = (ft^2/hr)
VA	Vapor velocity through active area (ft/sec)
z	Average weir length (ft)
Z	Liquid path (ft)
D	Tower diameter (ft)
W	Weir length (ft)
AA	Active area (ft^2)
G	Gas rate (pph)
L	Liquid rate (pph)
q	Liquid rate (ft^3/sec)
PG	Gas density (lb/ft^3)
PL	Liquid density (lb/ft^3)

VARIABLE	DESCRIPTION
m	Slope of equilibrium curve
E	Entrainment ratio-fractional
HL	Collapsed froth height (in.)
A,B	Calculated constants
PHI	Solvent association parameter
M1	Molecular weight—gases
M2	Molecular weight—gases
M	Molecular weight—liquid solute
V1	Molar volume—gases
V2	Molar volume—gases
V	Molar volume—liquid solute
P	Pressure (atm)
T	Temperature (K)
u_g	Gas viscosity (cp)

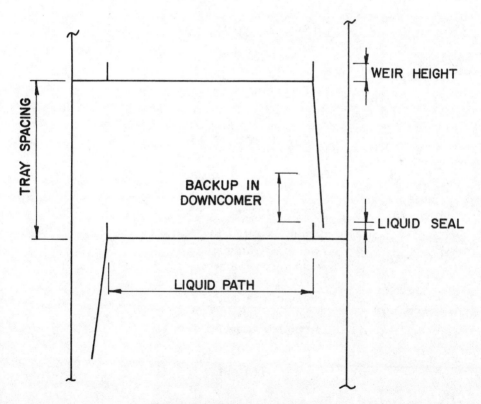

Figure 4.1 Tray nomenclature.

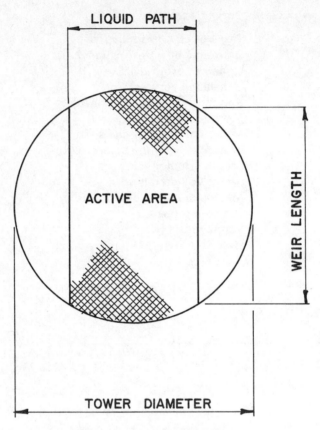

Figure 4.2 Tray Nomenclature.

VARIABLE	*DESCRIPTION*
u_l	Liquid viscosity (cp)
X1	Temporary variables
X2	Temporary variables
X3	Temporary variables
X4	Temporary variables
Y1	Temporary variables
Y2	Temporary variables

EXAMPLE

RUN

TRAY EFFICIENCY OF SIEVE PLATES

TOWER DIAMETER (FT) = ?   3

WEIR LENGTH (FT) = ?   2.15

VAPOR RATE (PPH) = ?   10000

LIQUID RATE (PPH) = ?   20000

GAS DENSITY (LB/FT3) = ?   .0363

LIQUID DENSITY (LB/FT3) = ?   60

GAS VISCOSITY (CP) = ?   .0125

LIQUID VISCOSITY (CP) = ?   .75

WEIR HT (IN) = ?   1

COLLAPSED FROTH HEIGHT(IN) = ?   .4315

LIQUID PATH LENGTH(FT) = ?   2.19

ENTRAINMENT RATIO(DEC) = ?   .0229

ACTIVE AREA(FT2) = ?   5.82

DO YOU HAVE GAS DIFFUSIVITY (Y/N):?   N

MOLE WT. OF COMPONENT 1 = ?   18

MOLE WT. OF COMPONENT 2 = ?   32

ATOMIC VOLUME OF COMPONENT 1 = ?   12.7

ATOMIC VOLUME OF COMPONENT 2 = ?   29.9

SYSTEM TEMPERATURE(K) = ?   373

SYSTEM PRESSURE(ATM) = ?   1

DO YOU HAVE LIQUID DIFFUSIVITY(Y/N) = :?   Y

LIQUID DIFFUSIVITY (SQR CM/SEC) = ?   1.6E-5

DO YOU HAVE SLOPE OF EQUIL CURVE (Y/N) = ?   N

RELATIVE VOLATILITY FOR SYSTEM = ?   4.05

SLOPE AT WHAT PT. ON CURVE(DEC) = ?   .10

TOWER DIAMETER(FT) = 3   WEIR LENGTH (FT) = 2.15

GAS RATE (PPH) = 10000   LIQUID RATE (PPH) = 20000

GAS DENSITY(LB/FT3) = .0363   LIQUID DENSITY (LB/FT3) = ; **60**

GAS VISCOSITY (CP) = .0125   LIQ. VISCOSITY (CP) = .75

DG(FT2/HR) = 1.2268   DL(FT2/HR) = 6.208E-05

WEIR HT (IN) = 1   COLLAPSED LIQ. HT(IN) = .4315

LIQUID PATH(FT) = 2.19   ACTIVE AREA (FT2) = 5.82

ENTRAINMENT RATIO = 2.29%   RELATIVE VOLATILITY = 4.05

SLOPE OF EQUIL CURVE = 2.37259

ESTIMATED TRAY EFFICIENCY = 52.0366%

?

TRAY EFF. AT ANOTHER POINT(Y/N)?   N

PROGRAM EXECUTION TERMINATED

READY

LISTING FOR BASIC PROGRAM "TRYEFF"

```
10 CLS

20 PRINT"TRAY EFFICIENCY OF SIEVE PLATES"

30 INPUT"TOWER DIAMETER(FT)=";D

40 INPUT"WEIR LENGTH(FT)=";W

50 INPUT"VAPOR RATE (PPH)=";G

60 INPUT"LIQUID RATE (PPH)=";L

70 INPUT"GAS DENSITY(LB/FT3)=";PG

80 INPUT"LIQUID DENSITY(LB/FT3)=";PL

90 PRINT

100 INPUT"GAS VISCOSITY(CP)=";VCP

110 INPUT"LIQUID VISCOSITY(CP)=";LCP

120 CLS

130 INPUT"WEIR HT(IN)=";HW
```

```
140 INPUT"COLLAPSED FROTH HEIGHT(IN)=";HL

150 INPUT"LIQUID PATH LENGTH(FT)=";Z1

160 INPUT"ENTRAINMENT RATIO(DEC)=";E

170 INPUT"ACTIVE AREA(FT2)=";AA

180 CLS

190 INPUT"DO YOU HAVE GAS DIFFUSIVITY(Y/N):";A$

200 IF A$="N" THEN GOSUB 680

210 IF A$="Y" THEN INPUT"GAS DIFFUSIVTIY(SQR CM/SEC)=";DG

220 INPUT"DO YOU HAVE LIQUID DIFFUSIVITY(Y/N):";A$

230 IF A$="N" THEN GOSUB 810

240 IF A$="Y" THEN INPUT"LIQUID DIFFUSIVITY(SQR CM/SEC)=";DL

250 DL=3.88*DL

260 DG=DG*3.88

270 SC=VCP*2.42/(DG*PG)

280 Q=L/(PL*3600)

290 VA=G/(PG*AA*3600)

292 CLS

294 INPUT"DO YOU HAVE SLOPE OF EQUIL CURVE(Y/N)";A$

296 IF A$="Y" THEN INPUT"SLOPE=";M:GOTO 330

300 INPUT"RELATIVE VOLATILITY FOR SYSTEM=";ALPHA

310 INPUT"SLOPE AT WHAT PT. ON CURVE(DEC)=";X1

320 GOSUB 900

330 ZZ=(D+W)/2

340 THETA=2.31E-5*HL*Z1*ZZ/Q

350 NL=7.31E5*DL[.5*(.26*VA*PG[.5+.15)*THETA

360 NG=(.776+.116*HW-.290*VA*PG[.5+9.72*Q/Z1)/SC[.5

370 NT=1/(1/NG+M*G/(L*NL))

380 EOG=1-EXP(-NT)
```

```
390 DE=(.774+1.026*VA+67.2*Q/Z1+.900*HW)[2

400 PE=Z1[2/(DE*THETA)

410 N=PE/2*((1+4*M*G*EOG/(PE*L))[.5-1)

420 X2=1-EXP(-(N+PE))

430 X3=EXP(N)-1

440 X4=N+PE

450 Y1=X4*(1+X4/N)

460 Y2=N*(1+N/X4)

470 E1=(X2/Y1+X3/Y2)*EOG

480 EMG=E1/(1+E1*(E/(1-E)))

490 CLS

500 PRINT"TOWER DIAMETER(FT)=";D;TAB(30);"WEIR LENGTH(FT)=";W

510 PRINT"GAS RATE(PPH)=";G;TAB(30);"LIQUID RATE(PPH)=";L

520 PRINT"GAS DENSITY(LB/FT3)=";PG;TAB(30);"LIQUID DENSITY(LB/FT3)=;PL

530 PRINT"GAS VISCOSITY(CP)=";VCP;TAB(30);"LIQ. VISCOSITY(CP)="LCP

540 PRINT"DG(FT2/HR)=";DG;TAB(30);"DL(FT2/HR)=";DL

550 PRINT

560 PRINT"WEIR HT(IN)=";HW;TAB(30);"COLLAPSED LIQ.HT(IN)=";HL

570 PRINT"LIQUID PATH(FT)=";Z1;TAB(30);"ACTIVE AREA(FT2)=";AA

580 PRINT"ENTRAINMENT RATIO=";E*100;"%";TAB(30);"RELATIVE VOLATILITY="
    ;ALPHA

590 PRINT"SLOPE OF EQUIL. CURVE=";M

600 PRINT

610 PRINT"ESTIMATED TRAY EFFICIENCY=";EMG*100;"%"

620 INPUT A$

630 CLS

640 INPUT"TRAY EFF. AT ANOTHER POINT(Y/N)";A$

650 IF A$="Y" THEN CLS:GOTO 292
```

```
660 PRINT"PROGRAM EXECUTION TERMINATED"

670 END

680 REM-CALCULATE GAS DIFFUSIVITY

690 CLS

700 INPUT"MOLE WT. OF COMPONENT 1=";M1

710 INPUT"MOLE WT. OF COMPONENT 2=";M2

720 INPUT"ATOMIC VOLUME OF COMPONENT 1=";V1

730 INPUT"ATOMIC VOLUME OF COMPONENT 2=";V2

740 INPUT"SYSTEM TEMPERATURE(K)=";T

750 INPUT"SYSTEM PRESSURE(ATM)=";P

760 CLS

770 A=.001*T[1.75*(1/M1+1/M2)[.5

780 B=P*(V1[.333+V2[.333)[2

790 DG=A/B

800 RETURN

810 REM-CALCULATE LIQUID DIFFUSIVITY

820 CLS

830 INPUT"MOLE WT. OF SOLVENT=";M

840 INPUT"SOLVENT ASSOCIATION PARAMETER=";PHI

850 INPUT"ATOMIC VOLUME OF SOLUTE=";V

860 INPUT"TEMPERATURE(K)=";T

870 CLS

880 DL=7.4E-8*SQR(PHI*M)*T/(LCP*V[.6)

890 RETURN

900 REM-CALCULATE SLOPE OF EQUILIBRIUM CURVE

910 GOSUB 970

920 F0=F

930 X1=X1+.001
```

```
940 GOSUB 970

950 M=(F-F0)/.001

960 RETURN

970 F=(ALPHA*X1)/(1+X1*(ALPHA-1))

980 RETURN
```

PACKED TOWER DESIGN

TOWER is a program used to calculate the diameter of packed distillation towers or packed absorption and stripping towers. The program is self-documenting, and you supply the following information:

1. Gas flow rate (pph)
2. Liquid flow rate (pph)
3. Temperature (F)
4. Pressure (psia)
5. Liquid density (lb/ft^3)
6. Liquid viscosity (cp)
7. Molecular weight of gas
8. Pressure drop per foot of packing (in. H$_2$O)
9. Tower Packing Factor

The program will immediately calculate the inside diameter of the tower. You may change any of the variables to obtain a different tower diameter. You do not have to reenter data that will not be changed. Simply press ENTER to proceed to the next variable, or begin the next calculation. Packing factors for some popular packings are listed in Table 4-3.

Equations for TOWER

$G1 = sqr(X1) * 3600$
$X1 = Y * DL * DG * 32.2 * CP^{.2}/CF * (DL/62.3)$
$Y = A + B * LN(X)$
$A = .02368 * DP^{.65278}$
$B = -.0241 * DP^{.91939}$
$DG = MW/V$

Perry, R., *Chemical Engineers' Handbook,* 5th Ed., McGraw Hill, New York, NY, 1973, pp. 18-23.
Design Information for Packed Towers, Norton Company, Akron, OH., 1980.

Table 4.3. TOWER PACKING FACTORS

PACKING TYPE	MAT'L.	NOMINAL PACKING SIZE (INCHES)									
		¼	⅜	½	⅝	¾	1 OR #1	1¼	1½	2 OR #2	3	3½ OR #3
Hy-Pak™	Metal						43			18		15
Super Intalox® Saddles	Ceramic						60			30		
Super Intalox Saddles	Plastic						33			21		16
Pall Rings	Plastic				97		52		40	24		16
Pall Rings	Metal				70		48		33	20		16
Intalox® Saddles	Ceramic	725	330	200		145	92		52	40	22	
Raschig Rings	Ceramic	1600	1000	580	380	255	155	125	95	65	37	
Raschig Rings	1/32" metal	700	390	300	170	155	115					
Raschig Rings	1/16" metal			410	290	220	137	110	83	57	32	
Berl Saddles*	Ceramic	900		240		170	110		65	45		

From *Design Information for Packed Towers*, Norton Company, Akron, OH., 1980.

V = [(T + 460)/492 * 14.7/P] * 359
A = G/G1
D1 = sqr(A/3.1416) * 24

Nomenclature for TOWER

VARIABLE	DESCRIPTION
G1	Flooding rate (lb/sec-ft^2)
X1	Temporary variable
X	Temporary variable
Y	Temporary variable
A	Correlation Constant
B	Correlation Constant
DG	Gas density (lb/ft^3)
DL	Liquid density (lb/ft^3)
MW	Gas molecular weight
V	Gas volume per mole (ft^3)
P	Pressure (psia)
A	Tower area (ft^2)
G	Gas rate (lb/hr)
L	Liquid rate (lb/hr)
D1	Tower diameter (ft)
CF	Packing factor
CP	Liquid viscosity (cp)
DP	Tower pressure drop (in./ft)
T	Temperature °F

EXAMPLE

READY

RUN

PACKED TOWER DESIGN

SYSTEM TITLE:?   **EXAMPLE**

VAPOR DATA

FLOW (PPH)=?   **10000**

TEMP(F)=?   **100**

PRESS(PSIA)=?   **14.7**

MOLE WT.=?   **29**

LIQUID DATA

FLOW(PPH)=? **100000**

DENSITY (LB/FT3)=? **62.3**

LIQUID VISCOSITY(CP)=? **1**

TOWER PARAMETERS

TYPE OF PACKING:= **2 " SADDLES**

PACKING FACTOR=? **21**

D/P PER FT. (IN.H2O)=? **.5**

SYSTEM TITLE:EXAMPLE

VAPOR DATA	LIQUID DATA
FLOW(PPH)=10000	100000
DENSITY(LB/FT3)=.066355	62.3
VISCOSITY"(CP)=------	1
MOLE WT.=29	------

TOWER PARAMETERS

PACKING TYPE:2" SADDLES   PACKING FACTOR=21

DP/FT(IN. H2O)=.5

CALC. DIA(INCHES)=34.37

PRESS (ENTER) TO CONTINUE?

DO YOU WISH TO CHANGE THE DATA (Y/N)? N

PROGRAM EXECUTION TERMINATED

READY

```
    LISTING FOR BASIC PROGRAM "TOWER"


 10 CLS
 20 N=0
 30 PRINT"PACKED TOWER DESIGN"
 40 INPUT"SYSTEM TITLE:";T$:CLS
```

```
 50 PRINT"VAPOR DATA"

 60 INPUT"FLOW (PPH)=";G

 70 INPUT"TEMP(F)=";T

 80 INPUT"PRESS(PSIA)=";P

 90 INPUT"MOLE WT.=";MW

100 IF N=1 THEN 220

110 CLS

120 PRINT"LIQUID DATA"

130 INPUT"FLOW(PPH)=";L

140 INPUT"DENSITY(LB/FT3)=";DL

150 INPUT"LIQUID VISCOSITY (CP)=";CP

160 IF N=2 THEN 220

170 CLS

180 PRINT"TOWER PARAMETERS"

190 INPUT"TYPE OF PACKING:";A$

200 INPUT"PACKING FACTOR=";CF

210 INPUT"D/P PER FT.(IN.H2O)=";DP

220 CLS

230 A=.02368*DP[.65278

240 B=-.0241*DP[.91939

250 V=(T+460)/460*14.7/P*359

260 DG=MW/V

270 X=L/G*SQR(DG/DL)

280 Y=A+B*LOG(X)

290 X1=Y*DL*DG*32.2*CP[.2/CF*(DL/62.3)

300 G1=SQR(X1)*3600:A=G/G1

310 D1=SQR(A/3.14126)*24

320 PRINT"SYSTEM TITLE:";T$
```

```
330 PRINT

340 PRINT TAB(10);"VAPOR DATA";TAB(40);"LIQUID DATA"

350 PRINT"FLOW(PPH)=";G;TAB(40);L

360 PRINT"DENSITY(LB/FT3)=";DG;TAB(40);DL

370 PRINT"VISCOSITY (CP)=";"  ------";TAB(40);CP

380 PRINT"MOLE WT.=";MW;TAB(40);"------"

390 PRINT

400 PRINT"TOWER PARAMETERS"

410 PRINT"PACKING TYPE:";A$;TAB(40);"PACKING FACTOR=";CF

420 PRINT"DP/FT(IN. H2O)=";DP

430 PRINT"CALC. DIA(INCHES)=";D1

440 INPUT"PRESS (ENTER) TO CONTINUE";X$

450 CLS

460 INPUT"DO YOU WISH TO CHANGE THE DATA(Y/N)";X$

470 IF X$="N" THEN 550

480 PRINT"1. VAPOR DATA"

490 PRINT"2. LIQUID DATA"

500 PRINT"3. TOWER PARAMETERS"

510 PRINT

520 INPUT"CHANGE WHICH VARIABLES (0-3)";N

530 IF N=0 THEN 550

540 ON N GOTO 50   ,120  ,180

550 PRINT"PROGRAM EXECUTION TERMINATED"

560 END
```

SIEVE PLATE HYDRAULICS

The design of sieve plate distillation or absorption towers can be an arduous task. There are various flooding correlations available, and most are too complex to be handled with programmable calculators. TRAY is a comprehensive sieve plate design program. It calculates the diameter of the sieve plate tower and a number of other important variables. You may change the tray geometry or the tray hydraulics, and the computer will recalculate a new tower diameter or new tray performance based on the existing tower diameter.

The program is self-documenting and asks for all data in common engineering units. The program calculates the collapsed froth height (inches of liquid) and the liquid path length, assuming that the downcomers take up 17.6% of the total tower area. Fig. 4.3 and Fig. 4.4 represent a typical sieve plate tower that can be designed using TRAY.

Equations for TRAY

$$A = .0062 * T + .0385$$
$$B = .00253 * T + .05$$

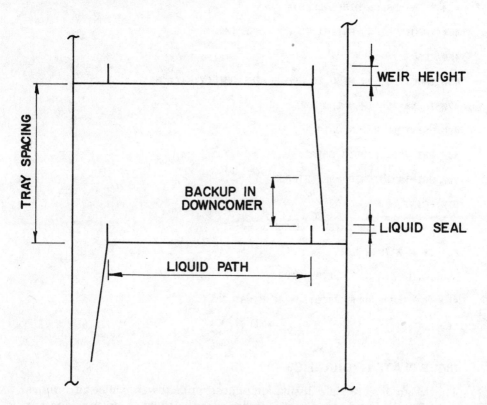

Figure 4.3 Tray Nomenclature.

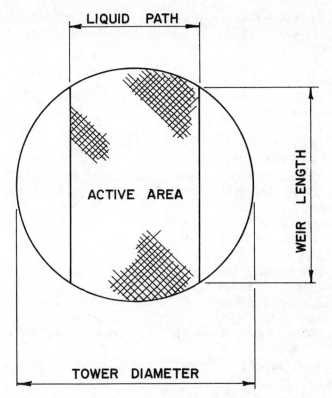

Figure 4.4 Tray Nomenclature.

$$X1 = A * \ln\{1/[(L/G) * sqr(PG/PL)]\}/2.303 + B$$
$$X2 = (SIGMA/20)^{.20} * (5 * X + .4)$$
$$CF = X1 * X2$$
$$VF = CF * sqr[(PL - PG)/PG]$$

For square Pitch
$$AA = P^2$$
$$AH = DH^2/4 * \pi$$

For triangular Pitch
$$AA = .5 * P^2 * .866$$
$$AH = DH^2/8 * \pi$$
$$X = AH/AA$$
$$V = VF * FR$$

$QG = G/(3600 * PG)$

$QL = L/(3600 * PL)$

$AN = QG/V$

$AT = AN/.912$

$D = sqr(4 * AT/\pi)$

$W = .7 * D$

$AD = .088 * AT$

$VA = V * AN/AA$

$VH = QG/(AA * X)$

$C\phi = 1.09 * (DH/L1)^{.25}$

$HD = 12 * C\phi * PG/PL * .40 * (1.25 - X) + (1 - X)^2 * VH^2/64.4$

$HL = .24 + .725 * HW - .29 * HW * VA * sqr(PG) + 4.48 *$
  $QL/Z1$

$Z1 = (D + W)/2$

$HR = .06 * SIGMA/(PL * DH)$

$Z = .71428 * D$

$HG = HD + HL + HR$

$AWS = (HW - LS) * W/12$

$H2 = .558 * (QL/AWS)^2$

$H1 = 5.385 * (QL/W)^{.667}$

$A1 = .06035 * FR - .16007 * FR^2 + .19488 * FR^3 - .07816 *$
  $FR^4 - .006879$

$B1 = -1.41719 * FR^{.975193}$

$E = A1 * X3^{B1}$

$X3 = L/G * sqr(PG/PL)$

Nomenclature for TRAY

VARIABLE	DESCRIPTION
A	Sieve tray constant
B	Sieve tray constant
T	Tray spacing (in.)
L	Liquid rate (pph)
G	Gas rate (pph)
PL	Liquid density (lb/ft^3)
PG	Gas density (lb/ft^3)
SIGMA	Surface tension (dyn/cm)
AN	Net tower area (ft^2)
AT	Tower area (ft^2)

Treyball, R., *Mass Transfer Operations,* 2nd. Ed., McGraw Hill, New York, NY, 1969, pp. 138–145.

VARIABLE	*DESCRIPTION*
CF	Tower flooding factor
VF	Flooding velocity (ft/sec)
AA	Pitch area (in.^{2})
P	Hole Pitch (in.)
DH	Hole diameter (in.)
AH	Hole area (in.^{2})
X	Tray open area—fractional
π	3.1416
V	Superficial vapor velocity (ft/sec)
QG	Gas flow rate (ft^3/sec)
QL	Liquid flow rate (ft^3/sec)
AN	Net tower area (ft^2)
D	Tower diameter (ft)
W	Weir length (ft)
AD	Down spout area (ft^2)
VA	Gas velocity through active area (ft/sec)
VH	Gas velocity through holes (ft/sec)
Cϕ	Orifice dicharge coefficient
HD	Dry gas pressure drop (in. of liquid)
HL	Collapsed liquid height (in.)
Z	Liquid path (ft)
HG	Tray pressure drop (in. of liquid)
HR	Residual pressure drop (in. of liq.)
AWS	Liquid seal flow area (ft^2)
H2	Liquid seal pressure drop (in. of liquid)
H1	Liquid crest over weir (in.)
A1	Entrainment constant
B1	Entrainment constant
E	Entrainment ratio-fractional
X3	Temporary variable
Z1	Average weir length (ft)
L1	Plate thickness (in.)

EXAMPLE

RUN

SIEVE TRAY HYDRAULICS

SYSTEM ID=?   **CCL4 COLUMN**

OLD VAPOR RATE (PPH)=0

VAPOR FLOW RATE (PPH)=?   **10000**

OLD VAPOR DENSITY (LT/FT3)=0
VAPOR DENSITY (LB/FT3)=?   **.3355**

OLD LIQUID RATE (PPH)=0
LIQUID FLOW RATE (PPH)=?   **10000**

OLD LIQUID DENSITY (LB/FT3)=0
LIQUID DENSITY (LB/FT3)=?   **99.55**

OLD SURFACE TENSION (DYNES/CM)=0
SURFACE TENSION (DYNES/CM)=?   **20**

TRAY LAYOUT=
PITCH TRIANGULAR(T), SQUARE(S)=?   **T**

OLD PITCH(IN)=0
PITCH(IN)=?   **.625**

OLD HOLE DIA. (IN)=0
HOLE DIA.(IN)=?   **.1875**
OLD TRAY THK (IN)=0
TRAY THK(IN)=?   **.125**

OLD FLOODING RATE(0-1.0)=0
FLOODING RATE(0-1.0)=?   **.65**

OLD TRAY SPACING(IN)=0
TRAY SPACING(IN)=?   **24**

OLD WEIR HEIGHT(IN)=0
WEIR HEIGHT(IN)=?   **1**

OLD LIQUID SEAL DEPTH(IN)=0
LIQUID SEAL DEPTH(IN)=?   **.5**

SYSTEM ID = CCL4 COLUMN

	LIQUID	GAS
PPH	10000	10000
DENSITY(LB/FT3)	99.55	.3355
SURFACE TENSION(DYNES/CM)	20	----

TRAY PARAMETERS

TRAY DIA. (FT) = 1.95485	TRAY SPACING(IN) = 24
PERF. DIA. (IN) = .1875	LAYOUT = TRIANGULAR
PITCH(IN) = .625	TRAY THK. (IN) = .125
WEIR LENGTH (FT) = 1.36839	WEIR HT. (IN) = 1
LIQUID SEAL(IN) = .5	LIQUID CREST(IN) = .401877
HYDRAULIC HEAD-HL(IN) = .477884	LIQUID PATH-Z (FT) = 1.39631
VAPOR VEL. (FT/SEC) = 3.02478	% OF FLOOD = 65

OPEN AREA OF TRAY (%) = 8.16562

CALC. PRESS. DROP (IN H2O) = 3.53545

LIQUID BACKUP IN DOWNCOMER(IN) = 2.34618

ENTRAINMENT RATIO(%) = 1.9181

MAKE ANY CHANGES(Y/N)?   N

BYE

READY

```
   LISTING FOR BASIC PROGRAM "TRAY"

10 CLS:PRINT"SIEVE TRAY HYDRAULICS":PI=3.1416

20 INPUT"SYSTEM ID=";N$

30 GOSUB 1170

40 GOSUB 1200

50 GOSUB 1230
```

```
 60 GOSUB 1260

 70 GOSUB 1290

 80 GOSUB 1320

 90 GOSUB 1380

100 GOSUB 1410

110 GOSUB 1440

120 GOSUB 1470

130 GOSUB 1500

140 GOSUB 1530

150 GOSUB 1560

160 LG=(L/G)*SQR(PG/PL) ' BEGIN CALCULATIONS

170 GOSUB 1080 'CALCULATE OPEN AREAS

180 IF A2$="N" THEN 260

190 IF LG<.1 THEN LG=.10

200 A=.0062*T+.0385:B=.00253*T+.05

210 X1=A*LOG(1/LG)/2.303+B

220 X2=(SIGMA/20)[.20*(5*X+.5)

230 CF=X1*X2

240 VF=CF*SQR((PL-PG)/PG)

250 V=VF*FR

260 QG=G/(3600*PG):QL=L/(3600*PL)

270 AN=QG/V

280 IF A2$="N" THEN 320

290 AT=AN/.912

300 D=SQR(4*AT/PI)

310 W=.7*D

320 AD=.088*AT

330 AA=AT-2*AD
```

340 VA=V*AN/AA

350 VH=QG/(AA*X)

360 C0=1.09*(DH/L1)[.25

370 HD=12*C0*PG/PL*(.40*(1.25-X)+(1-X)[2)*VH[2/64.4

380 Z1=(D+W)/2

390 HL=.24+.725*HW-.29*HW*VA*SQR(PG)+4.48*QL/Z1

400 HR=.06*SIGMA/(PL*DH)

410 Z=.71428*D

420 HG=HD+HL+HR

430 AWS=(HW-LS)*W/12

440 H2=.558*(QL/AWS)[2

450 H1=5.385*(QL/W)[.6667

460 GOSUB 1010

470 CLS:PRINT"SYSTEM ID=";N$

480 PRINT,,"LIQUID","GAS"

490 PRINT"PPH",,L,G

500 PRINT"DENSITY(LB/FT3)",,PL,PG

510 PRINT"SURFACE TENSION(DYNES/CM)",SIGMA,"----"

520 INPUT Z$

530 CLS:PRINT"TRAY PARAMETERS"

540 PRINT

550 PRINT"TRAY DIA.(FT)=";D,"TRAY SPACING(IN)=";T

560 PRINT"PERF.DIA.(IN)=";DH,"LAYOUT=";A1$

570 PRINT"PITCH(IN)=";P,,"TRAY THK.(IN)=";L1

580 PRINT"WEIR LENGTH(FT)=";W,"WEIR HT.(IN)=";HW

590 PRINT"LIQUID SEAL(IN)=";LS,"LIQUID CREST(IN)=";H1

600 PRINT"HYDRAULIC HEAD-HL(IN)=";HL,"LIQUID PATH-Z(FT)=";Z

610 PRINT"VAPOR VEL.(FT/SEC)=";V,"% OF FLOOD=";FR*100

```
620 PRINT

630 PRINT"OPEN AREA OF TRAY(%)=";X*100

640 PRINT"CALC. PRESS. DROP(IN H2O)=";HG*PL/62.3

650 PRINT"LIQUID BACKUP IN DOWNCOMER(IN)=";HG+H2

660 PRINT"ENTRAINMENT RATIO(%)=";E*100

670 INPUT Z$

680 CLS

690 INPUT"MAKE ANY CHANGES(Y/N)";Z$

700 IF Z$()"Y" AND  Z$()"N" THEN 690

710 IF Z$="N" THEN 990

720 GOTO 900

730 CLS

740 PRINT"1. VAPOR RATE"

750 PRINT"2. VAPOR DENSITY"

760 PRINT"3. LIQUID FLOW RATE"

770 PRINT"4. LIQUID DENSITY"

780 PRINT"5. SURFACE TENSION"

790 PRINT"6. HOLE LAYOUT"

800 PRINT"7. HOLE PITCH"

810 PRINT"8. HOLE DIA."

820 PRINT"9. TRAY THICKNESS"

830 PRINT"10. FLOODING RATE"

840 PRINT"11. TRAY SPACING"

850 PRINT"12. WEIR HEIGHT"

860 PRINT"13. LIQUID SEAL DEPTH"

870 RETURN

900 GOSUB 730

910 INPUT"CHANGE WHICH VARIABLE(1-13, 0 TO END)=";Z
```

```
920 IF Z=0 THEN 960

930 ON Z GOSUB 1170 ,1200 ,1230 ,1260 ,1290 ,1320 ,1380 ,1410 ,1440 ,1
    470 ,1500 ,1530 ,1560

940 GOTO 900

950 REM-RETURN TO RECALCULATE

960 INPUT"DO YOU WANT TO RESIZE TRAY(Y/N)";A2$

970 IF A2$<>"Y" AND A2$<>"N" THEN 960

980 GOTO 160

990 PRINT"***BYE***"

1000 END

1010 REM- CALCULATE ENTRAINMENT

1020 X3=L/G*SQR(PG/PL)

1030 A1=.06035*FR-.167007*FR[2+.19488*FR[3-.07816*FR[4-.006879

1040 B1=-1.41719*FR[.975193

1050 E=A1*X3[B1

1070 RETURN

1080 IF A$="T" THEN 1130

1090 AA=P[2

1100 AH=DH[2/4*PI

1110 X=AH/AA

1120 GOTO 1160

1130 AA=.5*P[2*.866

1140 AH=DH[2/8*PI

1150 X=AH/AA

1160 RETURN

1170 CLS:PRINT"OLD VAPOR RATE(PPH)=";G

1180 INPUT"VAPOR FLOW RATE(PPH)=";G

1190 RETURN
```

```
1200 CLS:PRINT"OLD VAPOR DENSITY(LB/FT3)=";PG

1210 INPUT"VAPOR DENSITY(LB/FT3)=";PG

1220 RETURN

1230 CLS:PRINT"OLD LIQUID RATE(PPH)=";L

1240 INPUT"LIQUID FLOW RATE(PPH)=";L

1250 RETURN

1260 CLS:PRINT"OLD LIQUID DENSITY(LB/FT3)=";PL

1270 INPUT"LIQUID DENSITY(LB/FT3)=";PL

1280 RETURN

1290 CLS:PRINT"OLD SURFACE TENSION(DYNES/CM)=";SIGMA

1300 INPUT"SURFACE TENSION(DYNES/CM)=";SIGMA

1310 RETURN

1320 CLS:PRINT"TRAY LAYOUT=";A1$

1330 INPUT"PITCH TRIANGULAR(T), SQUARE(S)=";A$

1340 IF A$()"T"AND A$()"S" THEN 1330

1350 IF A$="T" THEN A1$="TRIANGULAR":GOTO 1370

1360 A1$="SQUARE"

1370 RETURN

1380 CLS:PRINT"OLD PITCH(IN)=";P

1390 INPUT"PITCH(IN)=";P

1400 RETURN

1410 CLS:PRINT"OLD HOLE DIA.(IN)=";DH

1420 INPUT"HOLE DIA.(IN)=";DH

1430 RETURN

1440 CLS:PRINT"OLD TRAY THK(IN)=";L1

1450 INPUT"TRAY THK(IN)=";L1

1460 RETURN
```

```
1470 CLS:PRINT"OLD FLOODING RATE(0-1.0)=";FR

1480 INPUT"FLOODING RATE(0-1.0)=";FR

1490 RETURN

1500 CLS:PRINT"OLD TRAY SPACING(IN)=";T

1510 INPUT"TRAY SPACING(IN)=";T

1520 RETURN

1530 CLS:PRINT"OLD WEIR HEIGHT(IN)=";HW

1540 INPUT"WEIR HEIGHT(IN)=";HW

1550 RETURN

1560 CLS:PRINT"OLD LIQUID SEAL DEPTH(IN)=";LS

1570 INPUT"LIQUID SEAL DEPTH(IN)=";LS

1580 RETURN
```

5
COST ESTIMATION AND ECONOMICS

FLOW SHEET ESTIMATER

Viola et al. have proposed a general estimating procedure to obtain ±20% plant estimates when only flow sheet information is available. ESTIMATE is a computerized version of Viola's method. The information needed to use the program is a flow sheet that has the material balance, process pressures, and material of construction. ESTIMATE is best illustrated by an example.

PROBLEM: Management is considering building a new dimethyldichlorosilane plant to provide feedstock to a new organosilane plant. The project engineering group's task is to develop a ±20% estimate in less than one week. The plant is to produce 25 MM lb per year of DiMeDi. The principal chemical reaction is

$$SI + 2CH_3Cl = CH_3SiCl_2CH_3$$

The reaction stoichiometry shows us that it takes .217 lb of silicon metal and .783 lb of methylchloride to make 1 lb of DiMeDi. Assuming a 95% yield, the input–output ratio is approximately 1.00/.95 = 1.057. The plant will operate at 5–6 psig. The major process operating steps are shown in Table 5.1.

In general, each unit operation is considered a major process step. The more complex the step, the higher the weighting factor might be. For example, gas compression or simple distillation would be considered a single-process step. A reactor system consisting of more than a one-batch reactor system is handled by taking the number of batch reactors and dividing by

---

Viola, J., *Chem. Eng.*, April 6, 1981, pp. 80–86.

Table 5.1. Major Process Steps of DiMeDi Production[a]

STEP ID	N	SOLIDS	PRESS	% 304LSS
Ore handling	1	Yes	0	0
Reaction/quench	2	Yes	5	0
Crude distillation	1	No	5	0
MTCS distillation	1	No	5	0
Waste disposal	2	Yes	5	0
Final distillation	1	No	5	0
Prod. storage	.5	No	5	0
CH_3Cl vaporization	1	No	20	1

[a]Total effective steps = 9.5.

2. A table of suggested process-weighting factors is presented in Table 5.2. The engineer must use his or her own judgment in each case.

The estimated battery limits cost for the hypothetical DiMeDi plant is $6.73 MM. The estimated cost does not include offsites. Offsites can boost the overall plant cost by 40–60% of the battery limits cost.

The program is self-documenting and allows you to make changes and additions and delete process steps.

EXAMPLE

READY

RUN

FLOW SHEET ESTIMATER

PROBLEM ID:?   **DiMeDi PLANT**

PLANT CAPACITY (M #)=?   **25000**

INPUT/OUTPUT RATIO=?   **1.075**

DATE (MM/DD/YY):?   **04/01/84**

CE PLANT COST INDEX=?   **318**

ENTER INFORMATION FOR MAJOR OPERATING

STEPS. . . . . ENTER 'END' TO STOP

ID OF MAJOR OPERATING STEP:?   **ORE HANDLING**

MULTIPLIER FOR MAJOR OPERATING STEP=?   **1**

OPERATING PRESSURE (PSIG)=?   **0**

Table 5.2. Process-Weighting Factor

UNIT OPERATION	WEIGHTING FACTOR
Mixing/blending	1
Grinding	1
Heating/cooling/quench	1
Vaporization	1
Absorption processes	
Continuous	1
Multistage	Stages/2
Leaching processes	
Continuous	1
Multistage	Stages/2
Distillation	1
Absorption (charcoal)	1
Ion exchange	.5–1
Drying (continuous)	1
Reaction	
Continuous single stage	1
Continuous multistage	Stages/2
Batch	Batch reactors/2
Reforming	1.5
Fermentation (continuous)	2
Fermentation (batch 2 or more)	1.5 * Batch reactors/2
Gas compression	1
Product Storage	.25–1
Evaporation	
Long tube (continuous)	1
Multieffect	No. of effects/2
Filtration	1
Centrifugation	1

% STAINLESS STEEL (0-1.0)=?   **0**

DOES STEP HANDLE SOLIDS (Y/N):?   **Y**

ID OF MAJOR OPERATING STEP:?   **REACTION/QUENCH**

MULTIPLIER FOR MAJOR OPERATING STEP=?   **2**

OPERATING PRESSURE (PSIG)=?   **5**

% STAINLESS STEEL (0-1.0)=?   **0**

DOES STEP HANDLE SOLIDS (Y/N):?   **Y**

ID OF MAJOR OPERATING STEP:?   **CRUDE DISTILLATION**

MULTIPLIER FOR MAJOR OPERATING STEP=?   **1**

OPERATING PRESSURE (PSIG)=?   **5**

% STAINLESS STEEL (0-1.0)=?   **0**

DOES STEP HANDLE SOLIDS (Y/N):?   **N**

ID OF MAJOR OPERATING STEP:?   **MTCS DISTILL**

MULTIPLIER FOR MAJOR OPERATING STEP=?   **1**

OPERATING PRESSURE (PSIG)=?   **5**

% STAINLESS STEEL (0-1.0)=?   **0**

DOES STEP HANDLE SOLIDS (Y/N):?   **N**

ID OF MAJOR OPERATING STEP:?   **WASTE DISPOSAL**

MULTIPLIER FOR MAJOR OPERATING STEP=?   **2**

OPERATING PRESSURE (PSIG)=?   **5**

% STAINLESS STEEL (0-1.0)=?   **0**

DOES STEP HANDLE SOLIDS (Y/N):?   **Y**

ID OF MAJOR OPERATING STEP:?   **FINAL DISTILLATION**

MULTIPLIER FOR MAJOR OPERATING STEP=?   **1**

OPERATING PRESSURE (PSIG)=?   **5**

% STAINLESS STEEL (0–1.0)=?   **0**

DOES STEP HANDLE SOLIDS (Y/N):?   **N**

ID OF MAJOR OPERATING STEPS:?   **PROD. STORAGE**

MULTIPLIER FOR MAJOR OPERATING STEP=?   **.5**

OPERATING PRESSURE (PSIG)=?   **5**

% STAINLESS STEEL (0-1.0)=?   **0**

DOES STEP HANDLE SOLIDS (Y/N):?   **N**

ID OF MAJOR OPERATING STEP:?   **CH3CL VAPORIZATION**

MULTIPLIER FOR MAJOR OPERATING STEP=?   **1**

OPERATING PRESSURE (PSIG)=?   **30**

% STAINLESS STEEL (0-1.0)=?   **1**

DOES STEP HANDLE SOLIDS (Y/N):?   **N**

ID OF MAJOR OPERATING STEP:?   **END**

PRINTING RESULTS NOW

FLOW SHEET ESTIMATER

DATE:04/01/84   CE INDEX = 318

PROBLEM ID: DIMEDI PLANT   PLANT CAPACITY (M #) = 25000

I/O RATIO = 1.057

PROCESS STEP	N	PRESSURE	% 304L	SOLIDS
ORE HANDLING	1	0	0	YES
REACTION/QUENCH	2	5	0	YES
CRUDE DISTILLATION	1	5	0	NO
MTCS DISTILL	1	5	0	NO
WASTE DISPOSAL	2	5	0	YES
FINAL DISTILLATION	1	5	0	NO
PROD. STORAGE	.5	5	0	NO
CH3CL VAPORIZATION	1	30	100	NO

MAJOR PROCES STEPS = 8   N(EFF.) = 9.5

ESTIMATED PLANT COST = 6.73462E + 06

DO YOU WANT TO MAKE CHANGES(Y/N):?   **N**

PROGRAM EXECUTION TERMINATED

READY

```
    LISTING FOR BASIC PROGRAM "ESTIMATE"

 5 CLEAR 2000:CLS

10 DIM A$(20),N(20),P(20),MS(20),S(20),F$(20),S1(20)

20 PRINT"FLOW SHEET ESTIMATER"
```

```
30 INPUT"PROBLEM ID:";A$

40 INPUT"PLANT CAPACITY (M #)=";PC

50 INPUT"INPUT/OUTPUT RATIO=";IO

60 INPUT"DATE (MM/DD/YY):";D$

70 INPUT"CE PLANT COST INDEX=";CE

100 J=0

110 GOSUB 3000

220 LPRINT:LPRINT:F=0:S1=0:ST=0:N=0:S2=0

230 FOR I=1 TO J

235 IF A$(I)="DELETE" THEN GOTO 265

237 GOSUB 1000

240 N=N+N(I):ST=ST+1

245 S1=S1+S(I)*N(I):S2=S2+S1(I)*N(I)

250 IF A1$(I)="Y" THEN F=F+N(I):F$(I)="YES":GOTO 265

255 F$(I)="NO"

265 NEXT I

270 S=S1/N:FS=F/N:X1=IO/N

280 GOSUB 2000

290 CLS

300 PRINT"PRINTING RESULTS NOW"

310 LPRINT"FLOW SHEET ESTIMATER":LPRINT

320 LPRINT"DATE:";D$;TAB(40);"CE INDEX=";CE

330 LPRINT

340 LPRINT"PROBLEM ID:";A$;TAB(40);"PLANT CAPACITY (M #)=";PC

350 LPRINT

360 LPRINT"I/O RATIO=";IO

370 LPRINT
```

```
380 LPRINT"PROCESS STEP";TAB(30);"N";TAB(40);"PRESSURE";TAB(50);"% 304
    L";TAB(60);"SOLIDS"

385 LPRINT"-----------------------------------------------------------
    ------"

390 FOR I=1 TO J

395 IF A$(I)="DELETE" THEN 420

400 LPRINT A$(I);TAB(30);N(I);TAB(40);P(I);TAB(50);MS(I)*100;TAB(60);F
    $(I)

410 LPRINT

420 NEXT I

425 LPRINT"-----------------------------------------------------------
    ------"

430 LPRINT"MAJOR PROCESS STEPS=";ST;TAB(30);"N(EFF.)=";N

440 LPRINT

450 LPRINT"ESTIMATED PLANT COST=";COST

455 LPRINT CHR$(12)

460 INPUT"DO YOU WANT TO MAKE CHANGES(Y/N):";A2$

470 IF A2$="Y" THEN GOSUB 4000:GOTO 220

480 PRINT"PROGRAM EXECUTION TERMINATED"

490 END

1000 P=P(J):IF P(J)<100 THEN P=100

1005 B1=.5535*EXP(.00208*P)

1010 A1=.8833*EXP(.0014676*P)

1020 S(I)=A1+B1*MS(I)

1025 GOSUB 9000

1030 RETURN

2000 X2=.8795*EXP(.02991*X1)

2010 K=S*N*X2*(1-.6*FS)

2020 X3=(PC/20000)[.6*.394023*EXP(.146303*K)*1E6
```

```
2030 COST=X3*CE/305.6

2035 GOSUB 10000

2040 RETURN

3000 REM-DATA INPUT ROUTINE

3010 CLS

3020 PRINT"ENTER INFORMATION FOR MAJOR OPERATING STEPS...............ENT
     ER 'END' TO STOP"

3030 PRINT:J=J+1

3040 INPUT"ID OF MAJOR OPERATING STEP:";A$(J)

3050 IF A$(J)="END"THEN J=J-1:GOTO 3120

3060 INPUT"MULTIPLIER FOR MAJOR OPERATING STEP=";N(J)

3070 INPUT"OPERATING PRESSURE(PSIG)=";P(J)

3080 INPUT"% STAINLESS STEEL(0-1.0)=";MS(J)

3090 INPUT"DOES STEP HANDLE SOLIDS(Y/N):";A1$(J)

3100 CLS

3110 GOTO 3030

3120 RETURN

4000 REM- EDITING ROUTINE

4010 CLS

4020 PRINT "EDIT MODE"

4030 PRINT"1. ADD STEPS"

4040 PRINT"2. DELETE STEPS"

4050 PRINT"3. CHANGE STEPS"

4060 PRINT"WHICH FUNCTION(1-3, ENTER 0 TO END";

4070 INPUT N1

4080 IF N1=0 THEN RETURN

4090 ON N1 GOSUB 6000,7000,8000

4100 GOTO 4010
```

```
4500 CLS

4510 PRINT"ID OF MAJOR OPERATING STEP:";A$(J1)

4520 PRINT"MULTIPLIER FOR STEP=";N(J1)

4530 PRINT"OPERATING PRESSURE(PSIG)=";P(J1)

4540 PRINT"% STAINLESS STEEL(0-1.0)=";MS(J1)

4550 PRINT"HANDLING SOLIDS(Y/N):";A1$(J1)

4560 PRINT"-------------------------------------------------------------
     ----"

4570 INPUT"ID OF MAJOR OPERATING STEP:";A$(J1)

4580 INPUT"MULTIPLIER FOR MAJOR OPERATING STEP=";N(J1)

4590 INPUT"OPERATING PRESSURE(PSIG)=";P(J1)

4600 INPUT"% STAINLESS STEEL(0-1.0)=";MS(J1)

4610 INPUT"DOES STEP HANDLE SOLIDS(Y/N):";A1$(J1)

4620 RETURN

5000 CLS

5010 PRINT"STEP NO.","DESCRIPTION"

5020 FOR I1=1 TO J

5030 PRINT I1,A$(I1)

5040 NEXT I1

5050 RETURN

6000 CLS

6010 PRINT"EDIT MODE-ADDITIONS"

6020 GOSUB 3000

6030 RETURN

7000 CLS

7010 PRINT"EDIT MODE-DELETIONS"

7020 GOSUB 5010

7050 INPUT"DELETE WHICH STEP(ENTER 0 TO RETURN)";J1
```

```
7055 IF J1=0 THEN RETURN

7060 A$(J1)="DELETE"

7080 GOTO 7000

8000 CLS

8010 PRINT"EDIT MODE-STEP CHANGES"

8020 GOSUB 5010

8030 INPUT"CHANGE WHICH STEP(ENTER 0 TO END)";J1

8040 IF J1=0 THEN RETURN

8070 GOSUB 4500

8090 GOTO 8000

9000 M=.8+.00045*P+2.5E-6*P[2

9010 S1(I)=M*(MS(I)+.8)+.6

9020 RETURN

10000 C=PC*100:S3=S2/N

10010 Q=65*C[.6

10020 K=N*S3*X2*(1-.6*FS)

10030 Z1=(LOG(K/1.182)/2.303)[2.1

10040 Z2=CE*Q/304*10[Z1

10050 COST=(COST+Z2)/2

10060 RETURN
```

PRELIMINARY ECONOMICS—PROJECT SPECULATOR

SPEC is a program designed to evaluate the economic aspects of a proposed project and alternatives. The program allows you to forecast production costs, taking into account the effect of inflation and production expansion. The program takes into account the various factors shown in Table 5.3. All of these variables can be changed to allow you to examine alternatives.

The program prompts you for all required input. It assumes that all expenses and cash flows are continuous. The program calculates the discounted rate of return. You have the choice of printing a detailed report

Table 5.3. Economic Variables

VARIABLE COSTS	FIXED COSTS	DEPRECIATION	OTHER VARIABLES
Raw materials	Labor and supervision	Straight line	Production Product price
Power and energy	Maintenance	Declining balance	Growth rates Inflation Rates
Credits	Taxes and Insurance	Sum of the years' digits	
Miscellaneous	Miscellaneous		Project life (years) Capital cost

in which variable and fixed costs are broken down or just printing a summary.

By using ESTIMATE and SPEC, you can rapidly provide a very detailed analysis of project alternatives to company business managers. The two programs can also be used to estimate competitor capital expenses and cash flows.

When you choose to edit assumptions, you do not have to enter all of the data requested by the computer. You need only input the changes and skip over unchanged data by pressing ENTER when the computer prompts you for data.

The program takes several minutes to run, especially when the project life is 15 or more years. It can handle project lives as long as 20 years. If the project life is greater than 20 years, the DIM statements at line 10 must be changed to a value greater than 20.

EXAMPLE

RUN

PROJECT SPECULATOR

PROJECT TITLE=? **TEST**

VARIABLE COSTS

RAW MATERIAL COST($/UNIT PROD.)=0

INFLATION RATE(DEC)=0

RAW MATERIAL COST ($/UNIT PROD.)=? **.50**

INFLATION RATE (DEC)=? **.10**

VARIABLE COSTS

POWER AND ENERGY COST ($/UNIT PROD.)=   0

INFLATION RATE(DEC)=0

POWER AND ENERGY COST ($/UNIT PROD.)=?   .25

INFLATION RATE (DEC)=?   .15

VARIABLE COSTS

CREDITS ($/UNIT PROD)=0

INFLATION RATE (DEC)=0

CREDITS ($/UNIT PROD)=?   .05

INFLATION RATE (DEC)=?   .10

VARIABLE COSTS

MISC. EXPENSES ($/UNIT PROD.)=0

INFLATION RATE(DEC)=0

MISC. EXPENSES ($/UNIT PROD.)=?   .03

INFLATION RATE(DEC)=?   .06

FIXED COSTS

LABOR AND SUPERVISION=0

INFLATION RATE(DEC)=0

LABOR AND SUPERVISION=?   500000

INFLATION RATE(DEC)=?   .06

FIXED COSTS

MAINTENANCE=0

INFLATION RATE(DEC)=0

MAINTENANCE=?   500000

INFLATION RATE(DEC)=?   .06

FIXED COSTS

TAXES AND INSURANCE=0

TAXES AND INSURANCE=?　125000

FIXED COSTS

MISC FIXED COSTS=0

INFLATION RATE (DEC)=0

MISC FIXED COSTS=?　50000

INFLATION RATE(DEC)=?　.06

EXPECTED PRODUCTION(UNIT/YEAR)=0

INCOME/UNIT PROD.=0

GROWTH RATE(DEC)=0

INFLATION RATE(DEC)=0

EXPECTED PRODUCTION (UNIT/YEAR)=?　10000000

INCOME/UNIT PROD.=?　1.00

GROWTH RATE (DEC)=?　.05

INFLATION RATE(DEC)=?　.06

CAPITOL COST=0

SALVAGE VALUE=0

PROJECT LIFE(YEARS)=0

CAPITOL COST=?　5000000

SALVAGE VALUE=?　500000

PROJECT LIFE(YEARS)=?　10

DEPRECIATION SCHEDULE

1. STRAIGHT LINE

2. DECLINING BALANCE

3. SUM OF THE YEAR'S DIGITS

PRESENT SCHEDULE=

DEPRECIATION SCHEDULE (1-3)=?　1

TAX RATE(DEC)=0

TAX RATE (DEC)=? **.46**

ESTIMATED RATE OF RETURN=? **.25**

MAKE ANY CHANGES (Y/N)? **N**

PROJECT TITLE=TEST

YEAR	INC	VAR. COSTS	FIXED COSTS	DEPR	PBT	PAT	CASH FLOW
0	10,000,000	7,300,000	1,175,000	450,000	1,075,000	580,500	1,030,500
1	11,130,000	8,143,000	1,238,000	450,000	1,299,000	701,461	1,151,460
2	12,387,700	9,088,330	1,304,780	450,000	1,544,580	834,074	1,284,070
3	13,787,500	10,149,000	1,375,570	450,000	1,812,940	978,988	1,428,990
4	15,345,500	11,339,700	1,450,600	450,000	2,105,180	1,136,800	1,586,800
5	17,079,500	12,677,200	1,530,140	450,000	2,422,240	1,308,010	1,758,010
6	19,009,500	14,180,200	1,614,450	450,000	2,764,840	1,493,010	1,943,010
7	21,157,600	15,870,400	1,703,810	450,000	3,133,420	1,692,040	2,142,040
8	23,548,400	17,771,900	1,798,540	450,000	3,528,000	1,905,120	2,355,120
9	26,209,400	19,912,300	1,898,950	450,000	3,948,120	2,131,980	2,581,980

DISCOUNTED RATE OF RETURN = 25.8082%

FIXED CAPITOL=5E+06   WORKING CAPITOL=0
DEPRECIATION=STRAIGHT LINE   TAX RATE=46%
SALVAGE VALUE=500000   PROJECT LIFE=10 YEARS

ITEM	INFLATION RATE
$/UNIT PROD	6
GROWTH RATE	5
RAW MATERIALS	10
POWER AND ENERGY	15
CREDITS	10
MISC VAR.	6
LABOR+SUPV.	6
MAINTAINENCE	6
MISC FIXED	6

VARIABLE COST SUMMARY

YEAR	RAW MATR $/UNIT	POWER/ENERGY $/UNIT	CREDITS $/UNIT	MISC $/UNIT
0	0.5000	0.2500	0.0500	0.0300
1	0.5500	0.2875	0.0550	0.0318

(continued)

YEAR	RAW MATR $/UNIT	POWER/ENERGY $/UNIT	CREDITS $/UNIT	MISC $/UNIT
2	0.6050	0.3306	0.0605	0.0337
3	0.6655	0.3802	0.0666	0.0357
4	0.7321	0.4373	0.0732	0.0379
5	0.8053	0.5028	0.0805	0.0401
6	0.8858	0.5783	0.0886	0.0426
7	0.9744	0.6650	0.0974	0.0451
8	1.0718	0.7648	0.1072	0.0478
9	1.1790	0.8795	0.1179	0.0507

FIXED COST SUMMARY

YEAR	LABOR+SUPV $/UNIT	MAIN $/UNIT	TAXES+INS $/UNIT	MISC $/UNIT
0	0.0500	0.0500	0.0125	0.0050
1	0.0505	0.0505	0.0119	0.0050
2	0.0510	0.0510	0.0113	0.0051
3	0.0514	0.0514	0.0108	0.0051
4	0.0519	0.0519	0.0103	0.0052
5	0.0524	0.0524	0.0098	0.0052
6	0.0529	0.0529	0.0093	0.0053
7	0.0534	0.0534	0.0089	0.0053
8	0.0539	0.0539	0.0085	0.0054
9	0.0545	0.0545	0.0081	0.0054

SALES AND INCOME STATEMENT

YEAR	PROD	INCOME	$/UNIT
0	10,000,000	10,000,000	1.0000
1	10,500,000	11,130,000	1.0600
2	11,025,000	12,387,700	1.1236
3	11,576,300	13,787,500	1.1910
4	12,155,100	15,345,500	1.2625
5	12,762,800	17,079,500	1.3382
6	13,401,000	19,009,500	1.4185
7	14,071,000	21,157,600	1.5036
8	14,774,600	23,548,400	1.5938
9	15,513,300	26,209,400	1.6895

DEPRECIATION SCHEDULE
SCHEDULE-STRAIGHT LINE

YEAR	DEPR	VALUE ASSET	$/UNIT
0	450,000	4,550,000	0.0450
1	450,000	4,100,000	0.0429
2	450,000	3,650,000	0.0408
3	450,000	3,200,000	0.0389
4	450,000	2,750,000	0.0370
5	450,000	2,300,000	0.0353
6	450,000	1,850,000	0.0336
7	450,000	1,400,000	0.0320

YEAR	DEPR	VALUE ASSET	$/UNIT
8	450,000	950,000	0.0305
9	450,000	500,000	0.0290

MAKE ANY CHANGES (Y/N)?   **N**

PROGRAM EXECUTION TERMINATED

READY

```
    LISTING FOR BASIC PROGRAM "SPEC"

  5 REM-LPRINT CHR$(12) IS TOP OF FORM COMMAND FOR TRS-80 III

 10 D$(1)="STRAIGHT LINE":D$(2)="DECLINING BALANCE":D$(3)="SUM YEAR'S
    DIGITS":F$="###,###.###"

 20 VC$="VARIABLE COSTS":FC$="FIXED COSTS":INF$="INFLATION RATE(DEC)="

 30 CLS

 40 DIM VC(20),RM(20),PE(20),CR(20),MISC(20)

 50 DIM FC(20),LS(20),MN(20),MFC(20)

 60 DIM DCF(20),PBT(20),PAT(20),DPR(20)

 70 PRINT"PROJECT SPECULATOR"

 80 PRINT

 90 INPUT"PROJECT TITLE=";T$

100 GOSUB 770   'INPUT COST OF RAW MATERIALS

110 GOSUB 850   'INPUT COST OF POWER AND ENERGY

120 GOSUB 930   'INPUT CREDITS

130 GOSUB 1010  'INPUT MISC VAR. COSTS

140 GOSUB 1090  'INPUT LABOR AND SUPV.

150 GOSUB 1170  'INPUT MAINT. COSTS

160 GOSUB 1250  'INPUT TAXES AND INSUR.

170 GOSUB 1310  ' INPUT OTHER FIXED COSTS

180 GOSUB 1390  'INPUT FIRST YEAR'S SALES

190 GOSUB 1500  'INPUT CAPITAL COSTS AND SALVAGE VALUE
```

```
200 GOSUB 1590 'GET DEPRECIATION SCHEDULE

210 GOSUB 1690 'GET EFFECTIVE TAXE RATE

220 CLS

230 INPUT"ESTIMATED RATE OF RETURN=";RR

240 INPUT"MAKE ANY CHANGES(Y/N)";A$

250 IF A$="Y" THEN GOSUB 1990

260 REM-CALC DECLINING BALANCE FACTOR

270 F3=1-(SV/CT)[(1/N):TEMP=CT

280 REM-BEGIN CALCULATIONS

290 FOR I=0 TO N-1

300 GOSUB 1740 'CALCULATE VARIABLE COSTS

310 GOSUB 1810 'CALCULATE FIXED COSTS

320 GOSUB 1870 'CALCULATED DEPRECIATION

330 GOSUB 1960 'CALCULATE INCOME

340 PBT(I)=INC(I)-VC(I)-FC(I)-DPR(I)

350 PAT(I)=PBT(I)*(1-TXR)

360 IF PAT(I)<0 THEN PAT(I)=PBT(I)

370 CF(I)=PAT(I)+DPR(I)

380 NEXT I

390 REM-CALCULATE RATE OF RETURN

400 TEMP=RR

410 RR=RR+.0005

420 GOSUB 1990

430 F1=F

440 RR=RR-.0005

450 GOSUB 1990

460 F2=F
```

```
470 DERV=(F1-F2)/.0005

480 RR=RR-.50*F2/DERV

490 IF ABS((TEMP-RR)/RR)(=.001 THEN 520

500 TEMP=RR

510 GOTO 410

520 REM-PRINT RESULTS

530 LPRINT"PROJECT TITLE=";T$

540 LPRINT

550 LPRINT TAB(20);"VAR.";TAB(30);"FIXED";TAB(75);"CASH"

560 LPRINT"YEAR";TAB(9);"INC";TAB(20);"COSTS";TAB(30);"COSTS";TAB(43);
    "DEPR";TAB(54);"PBT";TAB(64);"PAT";TAB(74);"FLOW"

570 LPRINT

580 FOR I=0 TO N-1

590 LPRINT I;:LPRINT TAB(5);

600 LPRINT USING F$;INC(I);:LPRINT TAB(15);

610 LPRINT USING F$;VC(I);:LPRINTTAB(25);

620 LPRINT USING F$;FC(I);:LPRINT TAB(35);

630 LPRINT USING F$;DPR(I);:LPRINT TAB(45);

640 LPRINT USING F$;PBT(I);:LPRINT TAB(55);

650 LPRINT USING F$;PAT(I);:LPRINT TAB(65);

660 LPRINT USING F$;CF(I)

670 LPRINT

680 NEXT I

690 LPRINT"DISCOUNTED RATE OF RETURN=";RR*100;"%"

700 LPRINT:LPRINT:LPRINT

710 GOSUB 2260

720 CLS

730 INPUT"MAKE ANY CHANGES(Y/N)";A$
```

```
 740 IF A$="Y" THEN GOSUB 2060 :GOTO 260

 750 PRINT"PROGRAM EXECUTION TERMINATED"

 760 END

 770 CLS

 780 PRINT VC$

 790 PRINT"RAW MATERIAL COST($/UNIT PROD.)=";RM

 800 PRINT"INFLATION RATE(DEC)=";I1

 810 PRINT

 820 INPUT"RAW MATERIAL COST($/UNIT PROD.)=";RM

 830 INPUT"INFLATION RATE(DEC)=";I1

 840 RETURN

 850 CLS

 860 PRINT VC$

 870 PRINT"POWER AND ENERGY COST($/UNIT PROD.)=";PE

 880 PRINT"INFLATION RATE(DEC)=";I2

 890 PRINT

 900 INPUT"POWER AND ENERGY COST($/UNIT PROD.)=";PE

 910 INPUT"INFLATION RATE(DEC)=";I2

 920 RETURN

 930 CLS

 940 PRINT VC$

 950 PRINT"CREDITS($/UNIT PROD)=";CR

 960 PRINT INF$;I3

 970 PRINT

 980 INPUT"CREDITS($/UNIT PROD)=";CR

 990 PRINT INF$;:INPUT I3

1000 RETURN
```

```
1010 CLS

1020 PRINT VC$

1030 PRINT"MISC. EXPENSES($/UNIT PROD.)=";MISC

1040 PRINT INF$;I4

1050 PRINT

1060 INPUT"MISC. EXPENSES($/UNIT PROD.)=";MISC

1070 PRINT INF$;:INPUT I4

1080 RETURN

1090 CLS

1100 PRINT FC$

1110 PRINT"LABOR AND SUPERVISION=";LS

1120 PRINT INF$;I5

1130 PRINT

1140 INPUT"LABOR AND SUPERVISION=";LS

1150 PRINT INF$;:INPUT I5

1160 RETURN

1170 CLS

1180 PRINT FC$

1190 PRINT"MAINTENANCE=";MN

1200 PRINT INF$;I6

1210 PRINT

1220 INPUT"MAINTENANCE=";MN

1230 PRINT INF$;:INPUT I6

1240 RETURN

1250 CLS

1260 PRINT FC$

1270 PRINT"TAXES AND INSURANCE=";TI
```

```
1280 PRINT

1290 INPUT"TAXES AND INSURANCE=";TI

1300 RETURN

1310 CLS

1320 PRINT FC$

1330 PRINT"MISC FIXED COSTS=";MFC

1340 PRINT INF$;I7

1350 PRINT

1360 INPUT"MISC FIXED COSTS=";MFC

1370 PRINT INF$;:INPUT I7

1380 RETURN

1390 CLS

1400 PRINT"EXPECTED PRODUCTION(UNIT/YEAR)=";PROD

1410 PRINT"INCOME/UNIT PROD.=";NET

1420 PRINT"GROWTH RATE(DEC)=";I0

1430 PRINT"INFLATION RATE(DEC)=";I8

1440 PRINT

1450 INPUT"EXPECTED PRODUCTION (UNIT/YEAR)=";PROD

1460 INPUT"INCOME/UNIT PROD.=";NET

1470 INPUT"GROWTH RATE(DEC)=";I0

1480 INPUT"INFLATION RATE(DEC)=";I8

1490 RETURN

1500 CLS

1510 PRINT"CAPITAL COST=";CT

1520 PRINT"SALVAGE VALUE=";SV

1530 PRINT"PROJECT LIFE(YEARS)=";N

1540 PRINT

1550 INPUT"CAPITAL COST=";CT
```

```
1560 INPUT"SALVAGE VALUE=";SV

1570 INPUT"PROJECT LIFE(YEARS)=";N

1580 RETURN

1590 CLS

1600 PRINT"DEPRECIATION SCHEDULE"

1610 PRINT"1. STRAIGHT LINE"

1620 PRINT"2. DECLINING BALANCE"

1630 PRINT"3. SUM OF THE YEAR'S DIGITS"

1640 PRINT

1650 PRINT"PRESENT SCHEDULE=";D$(D)

1660 PRINT

1670 INPUT"DEPRECIATION SCHEDULE(1-3)=";D

1680 RETURN

1690 CLS

1700 PRINT"TAX RATE(DEC)=";TX

1710 PRINT

1720 INPUT"TAX RATE(DEC)=";TX

1730 RETURN

1740 REM-CALC VARIABLE COSTS

1750 RM(I)=RM*(1+I1)[I

1760 PE(I)=PE*(1+I2)[I

1770 CR(I)=CR*(1+I3)[I

1780 MISC(I)=MISC*(1+I4)[I

1790 VC(I)=(RM(I)+PE(I)+MISC(I)-CR(I))*PROD

1800 RETURN

1810 REM-CALC FIXED COSTS

1820 LS(I)=LS*(1+I5)[I

1830 MN(I)=MN*(1+I6)[I
```

```
1840 MFC(I)=MFC*(1+I7)[I

1850 FC(I)=LS(I)+MN(I)+MFC(I)+TI

1860 RETURN

1870 REM-CALC DEPRECIATION

1880 ON D GOTO 1890 ,1900 ,1940

1890 DPR(I)=(CT-SV)/N:RETURN

1900 VA=CT*(1-F3)[(I+1)

1910 DPR(I)=TEMP-VA

1920 TEMP=VA

1930 RETURN

1940 DPR(I)=(2*(N-(I+1)+1)/(N*(N+1)))*(CT-SV):VA=VA-DPR(I)

1950 RETURN

1960 REM-CALC INCOME

1970 INC(I)=NET*(1+I8)[I*PROD*(1+I0)[I

1980 RETURN

1990 REM-CALC DCF

2000 DCF=0

2010 FOR I=1 TO N

2020 DCF=DCF+CF(I-1)/(1+RR)[I

2030 NEXT I

2040 F=CT-DCF

2050 RETURN

2060 REM-EDITING ROUTINE

2070 CLS

2080 PRINT"1. RAW MATERIALS"

2090 PRINT"2. POWER AND ENERGY"

2100 PRINT"3. CREDITS"
```

```
2110 PRINT"4. MISC VARIABLE EXPENSES"

2120 PRINT"5. LABOR AND SUPERVISION"

2130 PRINT"6. MAINTENANCE"

2140 PRINT"7. TAXES AND INSURANCE"

2150 PRINT"8. MISC. FIXED COSTS"

2160 PRINT"9. 1ST YEAR INCOME"

2170 PRINT"10. CAPITAL COST, SALVAGE VALUE, PROJECT LIFE"

2180 PRINT"11. DEPRECIATION SCHEDULE"

2190 PRINT"12. EFFECTIVE TAX RATE"

2200 PRINT

2210 INPUT"CHANGE WHICH VARIABLE(0-12), ENTER 0 TO END";C

2220 IF C=0 THEN 2250

2230 ON C GOSUB 770 ,850  ,930  ,1010 ,1090 ,1170 ,1250 ,1310 ,1390 ,1
     500 ,1590 ,1690

2240 GOTO 2070

2250 RETURN

2260 REM-PRINT ASSUMPTIONS

2270 LPRINT"FIXED CAPITAL=";CT,"WORKING CAPITAL=";WC

2280 LPRINT"DEPRECIATION=";D$(D),"TAX RATE=";TX*100;"%"

2290 LPRINT"SALVAGE VALUE=";SV,"PROJECT LIFE=";N;"YEARS"

2300 LPRINT:LPRINT

2310 LPRINT"ITEM",,"INFLATION RATE"

2320 LPRINT"------------------------------------"

2330 LPRINT"$/UNIT PROD",,I8*100

2340 LPRINT"GROWTH RATE",,I0*100

2350 LPRINT"RAW MATERIALS",,I1*100

2360 LPRINT"POWER AND ENERGY",I2*100
```

```
2370 LPRINT"CREDITS",,I3*100

2380 LPRINT"MISC VAR.",,I4*100

2390 LPRINT"LABOR+SUPV.",,I5*100

2400 LPRINT"MAINTENANCE",,I6*100

2410 LPRINT"MISC FIXED",,I7*100

2420 LPRINT CHR$(12):F1$="###.####"

2430 LPRINT"VARIABLE COST SUMMARY"

2440 LPRINT

2450 LPRINT"YEAR","RAW MATR","POWER/ENERGY","CREDITS","MISC"

2460 LPRINT,"$/UNIT","$/UNIT","$/UNIT","$/UNIT"

2470 LPRINT

2480 FOR I=0 TO N-1

2490 LPRINT I,;

2500 LPRINT USING F1$;RM(I);:LPRINT,;

2510 LPRINT USING F1$;PE(I);:LPRINT,;

2520 LPRINT USING F1$;CR(I);:LPRINT,;

2530 LPRINT USING F1$;MISC(I)

2540 LPRINT

2550 NEXT I

2560 LPRINT CHR$(12)

2570 LPRINT"FIXED COST SUMMARY"

2580 LPRINT

2590 LPRINT"YEAR","LABOR+SUPV","MAIN","TAXES+INS","MISC"

2600 LPRINT,"$/UNIT","$/UNIT","$/UNIT","$/UNIT"

2610 LPRINT

2620 FOR I=0 TO N-1

2630 X=PROD*(1+I0)[I

2640 LPRINT I,;
```

```
2650 LPRINT USING F1$;LS(I)/X;:LPRINT ,;

2660 LPRINT USING F1$;MN(I)/X;:LPRINT ,;

2670 LPRINT USING F1$;TI/X;:LPRINT ,;

2680 LPRINT USING F1$;MFC(I)/X

2690 LPRINT

2700 NEXT I

2710 LPRINT CHR$(12)

2720 LPRINT"SALES AND INCOME STATEMENT"

2730 LPRINT

2740 LPRINT"YEAR","PROD","INCOME","$/UNIT"

2750 LPRINT

2760 FOR I=0 TO N-1

2770 X=PROD*(1+I0)[I

2780 Y=X*NET*(1+I8)[I

2790 LPRINT I,,;

2800 LPRINT USING F$;X;:LPRINT,;

2810 LPRINT USING F$;Y;:LPRINT,;

2820 LPRINT USING F1$;Y/X

2830 LPRINT

2840 NEXT I

2850 LPRINT CHR$(12)

2860 LPRINT"DEPRECIATION SCHEDULE"

2870 LPRINT"SCHEDULE=";D$(D)

2880 LPRINT

2890 VA=CT

2900 LPRINT"YEAR","DEPR","VALUE ASSET","$/UNIT"

2910 LPRINT

2920 FOR I=0 TO N-1
```

```
2930 X=PROD*(1+I0)[I

2940 VA=VA-DPR(I)

2950 LPRINT I;TAB(12);

2960 LPRINT USING F$;DPR(I);

2970 LPRINT TAB(30);:LPRINT USING F$;VA;

2980 LPRINT TAB(45);:LPRINT USING F1$;DPR(I)/X

2990 LPRINT

3000 NEXT I

3010 LPRINT CHR$(12)

3020 RETURN
```

ENGINEERING ECONOMICS—DISCOUNTED CASH FLOW ANALYSIS

To be a successful engineer, you must possess two talents. First, you must be technically competant in your chosen field; second, you must make sound economic decisions.

There are various ways of evaluating a project's profitability. Two common ways are the calculation of profitability indexes and of the discounted rate of return.

Profitability indexes are calculated by discounting the after tax cash flow at a fixed discount rate. The discounted cash flows are added up and divided by the project's capital cost. The resulting quotient is called the profitability index or PI. Numbers greater than one indicate that the project will generate cash flows that will meet the desired rate of return over a set number of years.

The discounted rate of return is determined by adding up the discounted cash flows at an assumed discount rate and comparing the total discounted cash flow to the capital cost of the project. When this difference is approximately zero, the discount rate at which this occurs is the rate of return for the project.

The program, DCF, will calculate the profitability index and the discounted rate of return for projects with even or uneven cash flows. The program prompts you for cash flows, prints the input data, allows you to change or correct input data, and then calculates the discounted rate of return and the profitability index. The program has a subroutine that will calculate depreciation schedules for the following depreciation methods:

1. Straight line
2. Declining balance
3. Sum of the year's digits

EXAMPLE

1. DISCOUNTED CASH FLOW

2. DEPRECIATION SCHEDULES

ENTER CHOICE (0-2), ENTER 0 TO END?   **2**

PROJECT TITLE=?   **TEST**

TOTAL CAPITAL INVESTMENT=?   **10000**

PROJECT LIFE(YRS)=?   **5**

1. STRAIGHT LINE

2. DECLINING BALANCE

3. SUM OF THE YEARS DIGITS

ENTER CHOICE (0-3), ENTER 0 TO END=?   **2**

CAPITAL INVESTMENT=10000

SALVAGE VALUE=?   **1000**

PROJECT TITLE=TEST

CAPITAL INVESTMENT=10000

SALVAGE VALUE=1000

PROJECT LIFE=5 YEARS

YEAR	DEPREC	ASSET VALUE
1	3690.43	6309.57
2	2328.5	3981.07
3	1469.19	2511.89
4	926.994	1584.89
5	584.894	1000

1. STRAIGHT LINE

2. DECLINING BALANCE

3. SUM OF THE YEARS DIGITS

ENTER CHOICE (0-3), ENTER 0 TO END =?   **0**

1. DISCOUNTED CASH FLOW

2. DEPRECIATION SCHEDULES

ENTER CHOICE (0-2), ENTER 0 TO END?   **0**

READY

EXAMPLE

1. DISCOUNTED CASH FLOW

2. DEPRECIATION SCHEDULES

ENTER CHOICE (0-2), ENTER 0 TO END?   **1**

PROJECT TITLE =?   **TEST**

TOTAL CAPITAL INVESTMENT =?   **10000**

PROJECT LIFE(YRS) =?   **5**

YEAR 1   CASH FLOW =?   **1000**

YEAR 2   CASH FLOW =?   **2000**

YEAR 3   CASH FLOW =?   **3000**

YEAR 4   CASH FLOW =?   **4000**

YEAR 5   CASH FLOW =?   **5000**

ESTIMATED RATE OF RETURN(DEC) =?   **.10**

PI HURDLE RATE(DEC) =?   **.10**

PROJECT TITLE = TEST

CAPITAL INVESTMENT = 10000

PROJECT LIFE = 5 YEARS

YEAR	CASH FLOW
1	1000
2	2000
3	3000
4	4000
5	5000

DISCOUNTED RATE OF RETURN(%) = 11.9973

PI VALUE = 1.0909

MAKE ANY CHANGES (Y/N)?   N

1. DISCOUNTED CASH FLOW

2. DEPRECIATION SCHEDULES

ENTER CHOICE (0-2), ENTER 0 TO END?   0

READY

```
   LISTING FOR BASIC PROGRAM "DCF"


10 DIM CF(20)

15 D$(1)="STRAIGHT LINE":D$(2)="DECLINING BALANCE":D$(3)="SUM OF THE
   YEAR'S DIGITS"

20 CLS

30 PRINT"1.DISCOUNTED CASH FLOW"

32 PRINT"2.DEPRECIATION SCHEDULES"

34 INPUT"ENTER CHOICE(0-2),ENTER 0 TO END";CH

36 IF CH=0 THEN 100

38 CLS

40 PRINT

50 INPUT"PROJECT TITLE=";T$

60 INPUT"TOTAL CAPITAL INVESTMENT=";CT
```

```
  70 INPUT"PROJECT LIFE(YRS)=";N

  80 ON CH GOSUB 1000,2000

  90 GOTO 20

 100 END

1000 CLS

1010 FOR I=1 TO N

1020 PRINT"YEAR";I;"   CASH FLOW=";

1030 INPUT CF(I)

1040 CLS

1050 NEXT I

1060 INPUT"ESTIMATED RATE OF RETURN(DEC)=";IRR

1065 INPUT"PI HURDLE RATE(DEC)=";I2

1100 REM-CALC RATE OF RETURN

1110 TEMP=IRR

1120 IRR=IRR+.0005

1130 GOSUB 1370

1140 F1=F

1150 IRR=IRR-.0005

1160 GOSUB 1370

1170 F2=F

1180 DERV=(F1-F2)/.0005

1190 IRR=IRR-.50*F2/DERV

1200 IF ABS((TEMP-IRR)/IRR)(=.001 THEN 1230

1210 TEMP=IRR

1220 GOTO 1120

1230 LPRINT"PROJECT TITLE=";T$

1240 LPRINT"CAPITAL INVESTMENT=";CT
```

```
1250 LPRINT"PROJECT LIFE=";N;"YEARS"

1260 LPRINT

1270 LPRINT"YEAR","CASH FLOW"

1280 LPRINT"-----------------------------------"

1285 DFV=0

1290 FOR I=1 TO N

1300 LPRINT I,CF(I):FV=CF(I)/(1+I2)[I

1305 DFV=DFV+FV

1310 NEXT I

1315 PI=DFV/CT

1320 LPRINT

1330 LPRINT"DISCOUNTED RATE OF RETURN(%)=";IRR*100

1335 LPRINT"PI VALUE=";PI

1340 LPRINT:LPRINT

1350 INPUT"MAKE ANY CHANGES(Y/N)";A$

1352 IF A$="Y" THEN GOSUB 3000:GOTO 1100

1354 RETURN

1360 CLS:GOTO 1070

1370 REM-CALC DCF

1380 DCF=0

1390 FOR I=1 TO N

1400 DCF=DCF+CF(I)/(1+IRR)[I

1410 NEXT I

1420 F=CT-DCF

1430 RETURN

2000 CLS

2010 PRINT"1. STRAIGHT LINE"
```

```
2020 PRINT"2. DECLINING BALANCE"

2030 PRINT"3. SUM OF THE YEAR'S DIGITS"

2040 INPUT"ENTER CHOICE(0-3), ENTER 0 TO END";D

2050 IF D=0 THEN RETURN

2060 CLS

2070 PRINT"CAPITAL INVESTMENT=";CT

2080 INPUT"SALVAGE VALUE=";SV

2090 CLS

2100 F3=1-(SV/CT)[(1/N)

2102 A=0

2120 VA=CT:TEMP=CT

2130 LPRINT"PROJECT TITLE=";T$

2140 LPRINT"CAPITAL INVESTMENT=";CT

2150 LPRINT"SALVAGE VALUE=";SV

2155 LPRINT"DEPR. SCHEDULE=";D$(D)

2160 LPRINT"PROJECT LIFE=";N;"YEARS"

2170 LPRINT

2172 LPRINT"YEAR","DEPREC","ASSET VALUE"

2180 FOR I=1 TO N

2190 ON D GOTO 2200,2210,2220

2200 DPR=(CT-SV)/N:VA=VA-DPR:GOTO 2230

2210 VA=CT*(1-F3)[I:DPR=TEMP-VA:TEMP=VA:GOTO 2230

2220 DPR=(2*(N-I+1)/(N*(N+1)))*(CT-SV):VA=VA-DPR

2230 LPRINT I,DPR,VA

2240 NEXT I

2245 LPRINT:LPRINT:LPRINT

2250 GOTO 2000

3000 CLS
```

```
3010 PRINT"EDITING ROUTINE"

3020 INPUT"LIST DATA TO LINE PRINTER(Y/N)";A$

3030 IF A$="N" THEN 3070

3040 FOR I=1 TO N

3050 LPRINT I,CF(I)

3060 NEXT I

3070 INPUT"HOW MANY CHANGES(ENTER 0 TO END)";C

3080 IF C=0 THEN 3170

3090 CLS

3100 FOR I=1 TO C

3110 INPUT"CHANGE WHICH VALUE";N1

3120 PRINT

3130 PRINT"OLD CASH FLOW=";CF(N1)

3140 INPUT"NEW CASH FLOW=";CF(N1)

3150 CLS

3160 NEXT I

3170 RETURN
```

TEMA HEAT EXCHANGER ESTIMATES

The chemical processing industries spend enormous sums of money on capital equipment such as heat exchangers. EXCOST is a program designed to provide ±15% estimates of TEMA heat exchangers. The program is based on the method outlined by G. P. Purohit.[1] It takes into account a wide variety of variables that affect the cost of tubular heat exchangers. The program is self-documenting and contains a data base for 52 common materials of construction. Table 5.4 lists the materials in the data base. The material factors in the data base reflect price ratios for welded and seamless tubing. Consult the original article for price factors of materials not in the data base.

Figure 5.1 is to be used to determine the type of TEMA heat exchanger

---

[1]Purohit, G. P., *Chem. Eng.,* August 22, 1983, pp. 56–67.

Table 5.4. Materials of Construction[2]

Carbon Steel	CS-1/2/MO
CS-3-1/2-NI	CS-2-1/2-NI
CS-1-MO	CS-2-NI-CU
304	304L
309	310
310L	316
316L	317
317L	321
329	330
347	405
410	430
439	444
446	904L
Ebrite	Carp-20-CB3
Carp-20-M06	Nickel 200
Monel 400	Inconel 600
Inconel 625	Incoloy 800
Incoloy 800H	Incoloy 825
Hastelloy C4	Hastelloy C276
Hastelloy G	Hastelloy X
Titanium II	Titanium VII
Titanium XII	90-10 Cupro
70-30 Cupro	Zirconium 702
Zirconium 705	Admiralty
Ferralium	Ebrite 26-1
Copper	Brass/navel

to be estimated. A single pass heat exchanger with fixed tubesheets and bonnet heads on both ends would be designated a TEMA Type BEM exchanger.

The program allows you to change materials of construction and have the computer recalculate a new price for the unit under consideration.

EXAMPLE

READY

RUN

TEMA EXCHANGER COST ESTIMATER

TEMA DESIGNATION FOR EXCHANGER = ?   **AEL**

SHELL DIA. (IN) = ?   **31**

NO. OF SHELL PASSES = ?   **1**

---

[2]Data for materials of construction excerpted from *Chemical Engineering* (Aug. 22) Copyright © 1983 by McGraw-Hill, Inc. New York, N.Y. 10020.

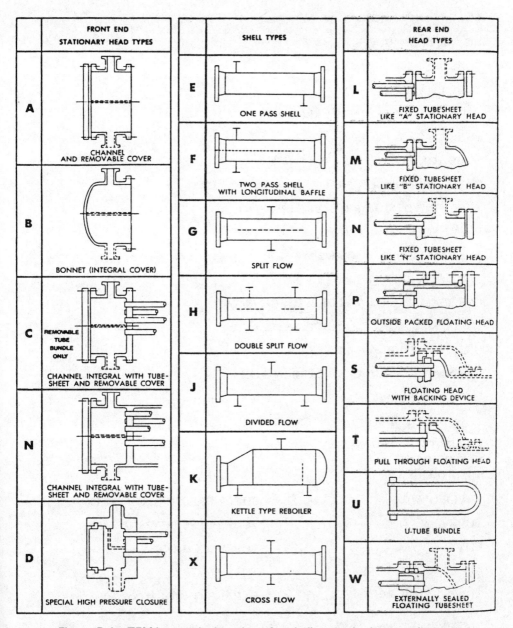

FRONT END STATIONARY HEAD TYPES	SHELL TYPES	REAR END HEAD TYPES
A CHANNEL AND REMOVABLE COVER	**E** ONE PASS SHELL	**L** FIXED TUBESHEET LIKE "A" STATIONARY HEAD
B BONNET (INTEGRAL COVER)	**F** TWO PASS SHELL WITH LONGITUDINAL BAFFLE	**M** FIXED TUBESHEET LIKE "B" STATIONARY HEAD
C REMOVABLE TUBE BUNDLE ONLY — CHANNEL INTEGRAL WITH TUBE-SHEET AND REMOVABLE COVER	**G** SPLIT FLOW	**N** FIXED TUBESHEET LIKE "N" STATIONARY HEAD
N CHANNEL INTEGRAL WITH TUBE-SHEET AND REMOVABLE COVER	**H** DOUBLE SPLIT FLOW	**P** OUTSIDE PACKED FLOATING HEAD
D SPECIAL HIGH PRESSURE CLOSURE	**J** DIVIDED FLOW	**S** FLOATING HEAD WITH BACKING DEVICE
	K KETTLE TYPE REBOILER	**T** PULL THROUGH FLOATING HEAD
	X CROSS FLOW	**U** U-TUBE BUNDLE
		W EXTERNALLY SEALED FLOATING TUBESHEET

Figure 5.1. TEMA-type designations for shell-and-tube heat exchangers. From *Standards of Tubular Exchanger Manufacturers Association,* 6th Ed., 1978. Reprinted with permission Tubular Exchanger Manufacturers Association Copyright © 1978.

TUBE LENGTH (FT)=?   16

NO. OF TUBE PASSES=?   4

TUBE OD. (IN)=?   .75

TUBE PITCH (IN)=?   1

SQUARE(S) OR TRIANGULAR(T) PITCH:?   T

TUBE BWG=?   12

TUBE COUNT/SHELL PASS=?   678

SHELL SIDE DESIGN PRESSURE (PSIG)=?   150

TUBE SIDE DESIGN PRESSURE (PSIG)=?   150

EXPANSION JOINT (Y/N):?   Y

TUBE MATERIAL:?   CS-1/2-MO

USE DATA BASE (Y/N):?   Y

USE SEAMLESS OR WELDED TUBING(S/W):?   S

SHELL MATERIAL:?   CARBON STEEL

USE DATA BASE (Y/N):?   Y

CHANNEL MATERIAL:?   CARBON STEEL

USE DATA BASE(Y/N):?   Y

TUBE SHEET MATERIAL:?   CS-1/2-MO

USE DATA BASE(Y/N):?   Y

TEMA TYEP:AEL

SHELL DIA(IN)=31	SHELL PRESSURE(PSIG)=150
TUBE LENGTH(FT)=16	TUBE PRESSURE(PSIG)=150
PITCH=T	TUBE PITCH(IN)=1
NO. OF TUBE PASSES=4	TUBE BWG=12
TUBE OD.(IN)=.75	EXPANSION JOINT(Y/N):Y

MATERIALS

TUBES:CS-1/2-MO   M1=2.6

SHELL:CARBON STEEL   M2 = 1

CHANNELS:CARBON STEEL   M3 = 1

TUBE SHEETS:CS-1/2-MO   M4 = 1.04

SURFACE AREA PER SHELL

(FT2) = 2130.01

NO. OF SHELLS = 1

ESTIMATED 1982 EXCHANGER

COST = 25188.1

CHANGE MATERIALS OF

CONSTRUCTION(Y/N) = ? N

PROGRAM EXECUTION TERMINATED

READY

```
    LISTING FOR BASIC PROGRAM "EXCOST"

 10 CLEAR 2000

 20 DIM MW(52),MS(52),M$(52),M(52)

 30 FOR I=1 TO 5

 40 READ A1$(I),F(I)

 50 NEXT I

 60 FOR I=1 TO 7

 70 READ A2$(I),CS(I)

 80 NEXT I

 90 FOR I=1 TO 8

100 READ A3$(I),R(I):NEXT I

110 FOR I=1 TO 52
```

```
120 READ M$(I),MW(I),MS(I),M(I)

130 NEXT I

140 CLS

150 PRINT"TEMA EXCHANGER COST ESTIMATER"

160 PRINT

170 INPUT"TEMA DESIGNATION FOR EXCHANGER=";A$

180 INPUT"SHELL DIA.(IN)=";DI

190 INPUT"NO. OF SHELL PASSES=";N

200 INPUT"TUBE LENGTH(FT)=";LT

210 INPUT"NO. OF TUBE PASSES=";NP

220 INPUT"TUBE OD.(IN)=";D0

230 INPUT"TUBE PITCH(IN)=";PI

240 INPUT"SQUARE(S) OR TRIANGULAR(T) PITCH:";P$

250 INPUT"TUBE BWG=";BWG

260 INPUT"TUBE COUNT/SHELL PASS=";NT

270 CLS

280 INPUT"SHELL SIDE DESIGN PRESSURE(PSIG)=";PS

290 INPUT"TUBE SIDE DESIGN PRESSURE(PSIG)=";PTS

300 INPUT"EXPANSION JOINT (Y/N):";X$

310 CLS

320 INPUT"TUBE MATERIAL:";T$

330 INPUT"USE DATA BASE(Y/N):";Z$

340 IF Z$="Y" THEN GOSUB 1090 :GOTO 360

350 INPUT"RATIO TUBE MAT'R/CS=";M1

360 INPUT"SHELL MATERTIAL:";S$

370 INPUT"USE DATA BASE(Y/N):";Z1$

380 IF Z1$="Y" THEN GOSUB 1180 :GOTO 400
```

```
390 INPUT"RATIO SHELL MAT'R/CS=";M2
400 INPUT"CHANNEL MATERIAL:";B$
410 INPUT"USE DATA BASE(Y/N):";ZZ$
420 IF ZZ$="Y" THEN GOSUB 1250 :GOTO 440
430 INPUT"RATIO CHANNEL MAT'R /CS=";M3
440 INPUT"TUBE SHEET MATERIAL:";TS$
450 INPUT"USE DATA BASE(Y/N):";Z3$
460 IF Z3$="Y" THEN GOSUB 1320 :GOTO 480
470 INPUT"RATIO TUBE SHEET MAT'R/CS=";M4
480 CLS
490 FOR I=1 TO 5
500 IF LEFT$(A$,1)=A1$(I) THEN F=F(I)
510 NEXT I
520 FOR I=1 TO 7
530 IF MID$(A$,2,1)=A2$(I) THEN CS=CS(I)
540 NEXT I
550 FOR I=1 TO 8
560 IF RIGHT$(A$,1)=A3$(I) THEN R=R(I)
570 NEXT I
580 REM- TEST FOR CERTAIN CONDITIONS
590 IF P$="T" THEN A=.85 ELSE A=1.0
600 IF X$="N" THEN CX=0:GOTO 620
610 CX=.467-.006655*DI-2.733E-5*DI[2
620 IF PS(=150 THEN C0S=0:GOTO 640
630 C0S=(PS/150-1)*(.07+.0016*(DI-12))
640 IF PTS(=150 THEN C1T=0:GOTO 660
650 C1T=(PTS/150-1)*(.035+.00056*(DI-12))
```

```
660 IF LT>=20 THEN CL=0:GOTO 680

670 CL=(1-LT/20)*(1.5-(.002083*(DI-12))/(1-LT/20))

680 Y=.045+.0016*(DI-12)*D0/.75*PI[2*A

690 G=1.4-.00714286*BWG

700 CG=Y*(G-1)

710 MT=Y*(M1-1)

720 MS=.1*(M2-1)

730 MC=.06*(M3-1)

740 TSM=.04*(M4-1)

750 IF NP<3 THEN CNP=0:GOTO 770

760 CNP=(NP-1)/100

770 P=.75*PI[2*A/D0

780 X=1-EXP((1-DI)/27)

790 B=6.6*P*F*R/X

800 A1=3.1416*D0*LT*NT/12*N

810 CT=CX+C0S+C1T+CL+CG+MT+MS+MC+TSM+CNP+CS

820 EB=B*(1+CT)*A1

830 CLS

840 LPRINT"TEMA TYPE:";A$

850 LPRINT"SHELL DIA(IN)=";DI;TAB(40);"SHELL PRESSURE(PSIG)=";PS

860 LPRINT"TUBE LENGTH(FT)=";LT;TAB(40);"TUBE PRESSURE(PSIG)=";PTS

870 LPRINT"PITCH=";P$;TAB(40);"TUBE PITCH(IN)=";PI

880 LPRINT"NO. OF TUBE PASSES=";NP;TAB(40);"TUBE BWG=";BWG

890 LPRINT"TUBE OD.(IN)=";D0;TAB(40);"EXPANSION JOINT(Y/N):";X$

900 LPRINT:LPRINT

910 LPRINT"MATERIALS"

920 LPRINT"--------------------------"

930 LPRINT"TUBES:";T$;TAB(40);"M1=";M1
```

```
940 LPRINT"SHELL:";S$;TAB(40);"M2=";M2

950 LPRINT"CHANNELS:";B$;TAB(40);"M3=";M3

960 LPRINT"TUBE SHEETS:";TS$;TAB(40);"M4=";M4

970 LPRINT"------------------------"

980 LPRINT"SURFACE AREA PER SHELL(FT2)=";A1/N

990 LPRINT"NO. OF SHELLS=";N

1000 LPRINT"------------------------"

1010 LPRINT"ESTIMATED 1982 EXCHANGER COST=";EB

1020 LPRINT:LPRINT

1030 LPRINT"PROGRAM EXECUTION TERMINATED"

1040 CLS

1050 INPUT"CHANGE MATERIALS OF CONSTRUCTION(Y/N):";Z4$

1060 IF Z4$="Y" THEN 310

1070 PRINT"PROGRAM EXECUTION TERMINATED"

1080 END

1090 REM-ROUTINE TO GET TUBE MATERIAL FACTORS

1100 INPUT"USE SEAMLESS OR WELDED TUBING(S/W):";T1$

1110 FOR I=1 TO 52

1120 IF T$=M$(I) THEN 1140

1130 GOTO 1160

1140 M1=MW(I)

1150 IF T1$="S" THEN M1=MS(I)

1160 NEXT I

1170 RETURN

1180 REM-ROUTINE TO GET SHELL MATERIAL FACTORS

1190 FOR I=1 TO 52

1200 IF S$=M$(I) THEN 1220

1210 GOTO 1230
```

```
1220 M2=M(I)

1230 NEXT I

1240 RETURN

1250 REM- ROUTINE TO GET CHANNEL FACTORS

1260 FOR I=1 TO 52

1270 IF B$=M$(I) THEN 1290

1280 GOTO 1300

1290 M3=M(I)

1300 NEXT I

1310 RETURN

1320 REM-ROUTINE TO GET MATERIAL FACTOR FOR TUBE FACE

1330 FOR I=1 TO 52

1340 IF TS$=M$(I) THEN 1360

1350 GOTO 1370

1360 M4=M(I)

1370 NEXT I

1380 RETURN

1390 DATA B,1,A,1.025,N,1.05,C,1.065,D,1.6

1400 DATA E,0,J,0,X,0,G,.075,H,.125,F,.175,K,.275

1410 DATA S,1,M,.8,L,.83,N,.85,U,.9,T,1.05,P,1.04,W,1.02

1420 DATA CARBON STEEL,1,2.5,1

1430 DATA CS-3-1/2-NI,1.20,3.10,1.20

1440 DATA CS-1-MO,1.05,2.70,1.05

1450 DATA 304,2.80,6.50,3.70

1460 DATA 309,5.80,14.5,7.70

1470 DATA 310L,7.60,12.40,10.10

1480 DATA 316L,4.80,11.00,6.40

1490 DATA 317L,8.30,13.60,8.30
```

```
1500 DATA 329,10.50,17.20,10.50

1510 DATA 347,5.50,13.70,7.30

1520 DATA 410,6.90,17.20,7.90

1530 DATA 439,5.00,11.20,5.80

1540 DATA 446,4.70,10.00,5.40

1550 DATA EBRITE,9.00,18.00,10.00

1560 DATA CARP 20-MO6,18.90,37.00,18.90

1570 DATA MONEL 400,15.50,15.50,14.50

1580 DATA INCONEL 625,16.35,32.70,27.40

1590 DATA INCOLOY 800H,18.00,18.00,20.00

1600 DATA HASTELLOY C4,28.70,40.00,31.30

1610 DATA HASTELLOY G,15.30,24.70,18.10

1620 DATA TITANIUM II,11.00,22.00,11.00

1630 DATA TITANIUM XII,14.00,28.00,14.00

1640 DATA 70-30 CUPRO,4.20,5.50,5.50

1650 DATA ZIRCONIUM 705,39.00,48.70,40.00

1660 DATA FERRALIUM,12.00,23.90,14.00

1670 DATA COPPER,4.20,4.20,4.20

1680 DATA CS-1/2-MO,1.04,2.60,1.04

1690 DATA CS-2-1/2-NI,1.15,2.90,1.15

1700 DATA CS-2-NI-CU,3.30,3.30,1.30

1710 DATA 304L,3.00,7.50,4.70

1720 DATA 310,7.40,12.00,9.80

1730 DATA 316,4.70,10.10,6.20

1740 DATA 317,8.10,13.30,8.10

1750 DATA 321,4.20,9.50,5.60

1760 DATA 330,7.90,12.90,9.50

1770 DATA 405,6.00,15.00,6.90
```

```
1780 DATA 430,5.40,10.60,6.20

1790 DATA 444,7.80,8.80,9.00

1800 DATA 904L,15.30,19.20,17.00

1810 DATA CARP-20-CB3,15.10,15.10,16.00

1820 DATA NICKEL 200,20.90,20.90,18.40

1830 DATA INCONEL 600,19.40,19.40,15.30

1840 DATA INCOLOY 800,11.00,21.80,9.00

1850 DATA INCOLOY 825,23.50,23.50,23.50

1860 DATA HASTELLOY C276, 29.10,38.10,31.00

1870 DATA HASTELLOY X,16.70,27.10,21.30

1880 DATA TITANIUM VII,21.00,42.00,21.00

1890 DATA 90-10 CUPRO,3.50,4.60,4.60

1900 DATA ZIRCONIUM 702,35.00,43.70,36.80

1910 DATA ADMIRALTY,3.60,3.60,3.60

1920 DATA EBRITE 26-1,9.00,9.00,10.00

1930 DATA BRASS/NAVEL,3.50,3.50,3.50
```

6
PHYSICAL PROPERTIES
OF PURE SUBSTANCES

PREDICTION OF PHYSICAL PROPERTIES

When projects are in the preliminary planning stage, estimates of chemical physical properties must be made because reliable data may not be available or may not be in the pressure and temperature ranges needed. The nine programs included in this chapter allow you to predict the following properties of pure compounds:

1. Critical temperature or TB/TC
2. Critical pressure
3. Critical volume
4. Critical compressibility factor
5. Acentric factor
6. Normal boiling point
7. Heat capacity of liquids and gases
8. Viscosity of gases
9. Viscosity of liquids
10. Density of liquids
11. Vapor pressure of liquids
12. Heat of vaporization for liquids
13. Thermal conductivity for gases
14. Thermal conductivity of nonpolar to slightly polar liquids
15. Surface tension of nonpolar to slightly polar liquids

The programs are designed so that you need only know the structure of the compound and the molecular weight. With this information, you should first determine the critical properties for the compound under consideration. The program PC is first used to determine the critical pressure. Next,

181

the program VC is used to estimate the critical volume. The program TC is used to determine the ratio of TB/TC. Then estimate the normal boiling point using option 1 of the program VAPOR. Once the normal boiling point is estimated, the critical temperature can be calculated.

With the above information known or estimated, the remaining physical properties such as viscosity, density, vapor pressure, and heat capacity can be calculated with a fair degree of accuracy.

CRITICAL TEMPERATURE ESTIMATES

The program TC is designed to calculate the critical temperature of a chemical compound when the normal boiling is known. If the normal boiling point is not known, the program returns the ratio of TB/TC that can be used in program VAPOR to estimate the normal boiling point after "VC" is used to calculate the critical volume.

The program determines the critical temperature or TB/TC ratio by summing contributions from various atomic and functional group values. The method used by the program is that proposed by Gambill.[1] The program has the symbols for each common chemical element and functional groups stored as data statements. The program prompts you to see if you wish to see the list. The symbols and their contributions are listed in Table 6.1. When entering atomic values for carbon, hydrogen, and nitrogen (except nitriles), count all atoms, not including those in the functional groups. As an example consider toluene. The contributions would be as follows:

SYMBOL	NO. PER MOLECULE
C	7
H	8
>C=C<	3
<6>	

For acetic acid the contributions would be as follows:

SYMBOL	NO. PER MOLECULE
C	1
H	3
R—COOH	1

Equations for TC

$$T_c = T_B/(\Sigma\Delta T/100)$$
$$\Delta T_{ALC} = 40.1693 * \text{Exp} (-.049922 * N1)$$

Table 6.1. Critical Temperature Symbols

SYMBOL	DESCRIPTION
C	Carbon
H	Hydrogen
N	Nitrogen (as amines)
CL	Chlorine
BR	Bromine
F	Fluorine
I	Iodine
S	Sulfur
SI	Silicon
R—OH	Alcohols
<O>—OH	Phenols
R—O—R	Ethers
>C=O	Ketones
R—COOH	Carboxyls
R—COOR	Esters
R—CN	Nitriles
>C=C<	Double bonds
R—CEC—R	Triple bonds
R—N<H2	Primary amines
R—N<R2	Secondary and higher amines
<5>	Five-membered rings
<6>	Six-membered rings
<N>	Nitrogen rings
BR2	Substitution of branching from second carbon
2BR2	Two substitutions or branching from second carbon
BR3	Substitution or branching from third carbon

Nomenclature for TC

VARIABLE	DESCRIPTION
T_C	Critical temperature (K)
T_B	Boiling point (K)
ΔT	Structural contribution
ΔT_{ALC}	Structural contributions for alcohols
N1	Number of carbons in alcohol

EXAMPLE

RUN"TC"

CRITICAL TEMPERATURE ESTIMATER

NAME OF COMPOUND=? **PROPIONITRILE**

NORMAL BOILING PT. (K)=? **351**

DO YOU WANT TO SEE THE LIST OF SYMBOLS(Y/N)? **N**

ENTER SYMBOLS FOR STRUCTURAL CONTRIBUTIONS

ENTER 'END' TO STOP

PRESS ENTER TO CONTINUE?

SYMBOL=? **C**

NO. OF UNITS PER MOLECULE=? **2**

SYMBOL=? **H**

NO. OF UNITS PER MOLECULE=? **5**

SYMBOL=? **R-CN**

NO. OF UNITS PER MOLECULE=? **1**

SYMBOL=? **END**

COMPOUND: PROPIONITRILE

SYMBOL	NO. PER MOLE	CONTRIBUTION
C	2	−55.32
H	5	28.52
R—CN	1	33.83

TB/TC=.6579

ESTIMATED CRITICAL TEMPERATURE(K)= 533.516

READY

LISTING FOR BASIC PROGRAM "TC"

```
10 DIM A1$(26),T(26),AA$(26),N(26),B$(26)

20 FOR I=1 TO 26

30 READ A1$(I),T(I)
```

```
40 NEXT I

50 FOR I=1 TO 26

60 READ B$(I)

70 NEXT I

80 CLS

90 PRINT"CRITICAL TEMPERATURE ESTIMATER"

95 INPUT"NAME OF COMPOUND=";C$

100 INPUT"NORMAL BOILING PT.(K)=";TB

110 INPUT"DO YOU WANT TO SEE THE LIST OF SYMBOLS(Y/N)";A$

120 IF A$="Y" THEN GOSUB 1000

130 CLS

140 N=1

150 PRINT"ENTER SYMBOLS FOR STRUCTURAL CONTRIBUTIONS"

160 PRINT"ENTER 'END' TO STOP"

170 INPUT"PRESS ENTER TO CONTINUE";Z$

180 CLS

190 FLAG=1

200 INPUT"SYMBOL=";AA$(N)

210 IF AA$(N)="END" THEN 320

220 IF AA$(N)="R-OH" THEN GOSUB 500

230 IF AA$(N)="BR2" OR AA$(N)="BR3"OR AA$(N)="2BR2" THEN N(N)=1:GOTO 3
    00

240 FOR I=1 TO 26

250 IF AA$(N)=A1$(I) THEN FLAG=0

260 NEXT I

270 IF FLAG=0 THEN 295

280 PRINT"SYMBOL NOT FOUND......TRY AGAIN"

285 INPUT "PRESS 'ENTER' TO CONTINUE";A$
```

```
290 GOTO 180

295 INPUT"NO. OF UNITS PER MOLECULE=";N(N)

300 N=N+1

310 GOTO 180

320 REM-BEGIN CALCULATIONS

330 FOR I=1 TO N-1

340 FOR J=1 TO 26

350 IF AA$(I)<>A1$(J) THEN 370

360 T=T+T(J)*N(I):T2(I)=T(J)

370 NEXT J

380 NEXT I

390 TC=TB/(T/100)

400 CLS

410 LPRINT"COMPOUND:";C$

415 LPRINT

420 LPRINT"SYMBOL","NO. PER MOLE","CONTRIBUTION"

425 LPRINT

430 FOR I=1 TO N-1

440 LPRINT AA$(I),N(I),T2(I)

450 NEXT I

460 LPRINT

465 LPRINT"TB/TC=";T/100

470 LPRINT"ESTIMATED CRITICAL TEMPERATURE(K)=";TC

480 END

490 REM-CALCULATE SPECIAL RELATIONSHIPS

500 REM-CALCULATE ALCOHOL CONTRIBUTION

510 INPUT"HOW MANY CARBON ATOMS IN COMPOUND=";N1

520 T(10)=40.1693*EXP(-.0409922*N1)
```

```
 530 RETURN

1000 LPRINT"SYMBOL","CONTRIBUTION","DESCRIPTION"

1010 FOR I=1 TO 26

1020 LPRINT A1$(I),T(I),B$(I)

1030 LPRINT

1040 NEXT I

1050 RETURN

2000 DATA C,-55.32,H,28.52,N,30.6,CL,29.89,BR,31.15,F,29.75,I,30.50,S,1
     .31,SI,-54.00

2010 DATA R-OH,0,(O)-OH,31.6,R-O-R,1.59,)C=O,58.96,R-COOH,35.94,R-COOR,
     4.12,R-CN,33.83,)C=C(,55.91,R-CEC-R,112.9,R-N(H2,-18.37,R-N(R2,-
     19.17,(5),54.28,(6),53.52,(N),-26.92,BR2,-0.34,2BR2,-1.42,BR3,-0
     .96

3000 DATA CARBON,HYDROGEN,NITROGEN (AS AMINES),CHLORINE,BROMINE,FLUORIN
     E,IODINE,SULFUR,SILICON

3010 DATA ALCOHOLS,PHENOLS,ETHERS,KETONES,CARBOXYLS,ESTERS,NITRILES,DOU
     BLE BONDS,TRIPLE BONDS,PRIMARY AMINES,SECONDARY AND HIGHER AMINE
     S,FIVE-MEMBERED RINGS,SIX-MEMBERED RINGS,NITROGEN RINGS

3020 DATA SUBSTITUTION OR BRANCHING FROM THE 2ND CARBON,TWO SUBSTITUTIO
     NS OR BRANCHES FROM THE 2ND CARBON,SUBSTITUTION OR BRANCHING FRO
     M THE 3RD CARBON
```

CRITICAL PRESSURE ESTIMATES

The program PC is used to predict the critical pressure of compounds. You must have the molecular weight and structure of the compound to use the program. The program calculates the critical pressure by summing structural contributions. The symbols and values for the structural contributions are contained in DATA statements at the end of the program. The program prompts you to enter the symbol for each contribution and the number per molecule. When entering atomic values for carbon, oxygen, nitrogen, and hydrogen, count only those atoms not included in functional groups. For example, the contributions for ethylacetate, $CH_3COOC_2H_5$, would be

SYMBOL	NO. PER MOLECULE
C	3
H	8
R—COOR	1

For a cyclic hydrocarbon such as cyclohexane, the contributions would be

SYMBOL	NO. PER MOLECULE
C	6
H	12
<6>	1

The program uses the contribution method suggested by Gambill.[2] Table 6.2 is a listing of the symbols and their descriptions.

Equations for PC

$$P_C = 10000 \; MW/\Sigma\Delta p_i^2$$

Table 6.2.   Critical Pressure Symbols

SYMBOL	DESCRIPTION
C	Carbon
H	Hydrogen
CL	Chlorine
BR	Bromine
F	Fluorine
I	Iodine
S	Sulfur
SI	Silicon
R—OH	Alcohols
<O>—OH	Phenols
R—O—R	Ethers
>C=O	Ketones
R—COOH	Carboxyls
R—NO2	Nitro group
R—COOR	Esters
R—CHO	Aldehydes
R—CN	Nitriles
>C=C<	Double bonds (Olefins)
R—CEC—R	Triple bonds
R—N<H2	Primary amines
R—N<R2	Secondary and higher amines
<5>	Five-membered rings (sat'd)
<6>	Six-membered rings (sat'd)
<0>	Benzene ring
BR2	Substitution or branching on second carbon
BR3	Substitution or branching on third carbon

Nomenclature for PC

VARIABLE	DESCRIPTION
P_C	Critical pressure (atm)
MW	Molecular weight
Δp_i	Structural contributions (Iodine $= 112.0$)

EXAMPLE

RUN

CRITICAL PRESSURE ESTIMATER

NAME OF COMPOUND = ? **PROPIONITRILE**

MOLECULAR WEIGHT = ? **69**

DO YOU WANT TO SEE THE LIST OF SYMBOLS(Y/N)? **N**

ENTER SYMBOLS FOR STRUCTURAL CONTRIBUTIONS

ENTER 'END' TO STOP

PRESS ENTER TO CONTINUE?

SYMBOL = ? **C**

NO. OF UNITS PER MOLECULE = ? **2**

SYMBOL = ? **H**

NO. OF UNITS PER MOLECULE = ? **5**

SYMBOL = ? **R-CN**

NO. OF UNITS PER MOLECULE = ? **1**

SYMBOL = ? **END**

COMPOUND: PROPIONITRILE

SYMBOL	NO. PER MOLE	CONTRIBUTION
C	2	-9.35
H	5	16.2
R—CN	1	52.5

ESTIMATED CRITICAL PRESSURE (ATM.) = 52.3558

READY

LISTING FOR BASIC PROGRAM "PC"

```
10 DIM A1$(25),P(25),AA$(25),N(25),B$(25)

20 FOR I=1 TO 25

30 READ A1$(I),P(I)

40 NEXT I

50 FOR I=1 TO 25

60 READ B$(I)

70 NEXT I

80 CLS

90 PRINT"CRITICAL PRESSURE ESTIMATER"

95 INPUT"NAME OF COMPOUND=";C$

100 INPUT"MOLECULAR WEIGHT=";MW

110 INPUT"DO YOU WANT TO SEE THE LIST OF SYMBOLS(Y/N)";A$

120 IF A$="Y" THEN GOSUB 1000

130 CLS

140 N=1

150 PRINT"ENTER SYMBOLS FOR STRUCTURAL CONTRIBUTIONS"

160 PRINT"ENTER 'END' TO STOP"

170 INPUT"PRESS ENTER TO CONTINUE";Z$

180 CLS

190 FLAG=1

200 INPUT"SYMBOL=";AA$(N)

210 IF AA$(N)="END" THEN 320

220 IF AA$(N)=")C=C(" THEN GOSUB 500:GOTO 300

230 IF AA$(N)="BR2" OR AA$(N)="BR3" THEN N(N)=1:GOTO 300

240 FOR I=1 TO 25

250 IF AA$(N)=A1$(I) THEN FLAG=0
```

```
260 NEXT I

270 IF FLAG=0 THEN 295

280 PRINT"SYMBOL NOT FOUND......TRY AGAIN"

285 INPUT "PRESS `ENTER` TO CONTINUE";A$

290 GOTO 180

295 INPUT"NO. OF UNITS PER MOLECULE=";N(N)

300 N=N+1

310 GOTO 180

320 REM-BEGIN CALCULATIONS

330 FOR I=1 TO N-1

340 FOR J=1 TO 25

350 IF AA$(I)<>A1$(J) THEN 370

360 P1=P1+N(I)*P(J):P2(I)=P(J)

370 NEXT J

380 NEXT I

390 PC=10E3*MW/P1[2

400 CLS

410 LPRINT"COMPOUND:";C$

415 LPRINT

420 LPRINT"SYMBOL","NO. PER MOLE","CONTRIBUTION"

425 LPRINT

430 FOR I=1 TO N-1

440 LPRINT AA$(I),N(I),P2(I)

450 NEXT I

460 LPRINT

470 LPRINT"ESTIMATED CRITICAL PRESSURE(ATM.)=";PC

480 END

490 REM-CALCULATE DOUBLE BOND CONTRIBUTIONS
```

```
500 INPUT"NO. OF CARBONS IN CHAIN CONTAINING THE DOUBLE BOND=";N1

510 X=23.4068+5.31082*N1-1.4821*N1[2+0.0669765*N1[3

520 P2(N)=X

530 P1=P1+X

540 N(N)=1

550 RETURN

1000 LPRINT"SYMBOL","CONTRIBUTION","DESCRIPTION"

1010 FOR I=1 TO 25

1020 LPRINT A1$(I),P(I),B$(I)

1030 LPRINT

1040 NEXT I

1050 RETURN

2000 DATA C,-9.35,H,16.20,CL,48.0,BR,68.8,F,39.9,S,27.8,SI,22.4

2010 DATA R-OH,23.7,<0>-OH,23.7,R-O-R,17.0,>C=O,30.2,R-COOH,51.7,R-NO2,
     60,R-COOR,47.5,R-CHO,50.40,R-CN,52.5,>C=C<,0,R-CEC-R,51.1,R-N<H2
     ,-3.15,R-N<R2,-3.15,<5>,10.5,<6>,7.2,<0>,84.5,BR2,-1.62,BR3,-4.7
   5

3000 DATA CARBON,HYDROGEN,CHLORINE,BROMINE,FLUORINE,SULFUR,SILICON

3010 DATA ALCOHOLS,PHENOLS,ETHERS,KETONES,CARBOXYLS,NITRO GROUP,ESTERS,
     ALDEHYDES,NITRILES,DOUBLE BONDS,TRIPLE BONDS,PRIMARY AMINES,SECO
     NDARY AND HIGHER AMINES

3020 DATA FIVE MEMBERED RINGS(SAT'D),SIX MEMBERED RINGS(SAT'D),BENZENE
     RING,SUBSTITUTION OR BRANCHING ON 2ND CARBON,SUBSTITUTION OR BRA
     NCHING ON 3RD CARBON
```

CRITICAL VOLUME ESTIMATES

Critical volumes of compounds may be estimated by the method recommended by Gold.[3] The program VC is a modified version of the method Vowles first proposed. The program determines the critical volume by summing the atom contributions and functional group contribution. Unlike the program PC, *all* atoms of each element are included in the summation. For example, the contributions for metachlorotoluene would be the following:

CONTRIBUTION	NO. PER MOLECULE
C	7
>C=C<	3
H	7
Cl	1
<O>	1

The contribution for a compound such as furfural $C_5H_4O_2$ would be the following:

SYMBOL	NO. PER MOLECULE
C	5
H	4
>C=O	1
>C=C<	2
R—O—R	1
<5>	1

Critical volumes may also be estimated by making use of the relationship.

$$Vc = \frac{ZcRTc}{Pc}$$

where R = 82.0567 and Zc is obtained by running option 1 with program "VAPOR". Structural symbols used by VC are listed in Table 6.3.

Equations for VC

$V_C = \Sigma \Delta V_i$

Nomenclature for VC

VARIABLE	DESCRIPTION
V_c	Critical volume (ml/mole)
ΔV_i	Structural contributions

EXAMPLE

READY

RUN

Table 6.3 Critical Volume Symbols

SYMBOL	DESCRIPTION
C	All carbon
H	All hydrogen
O	All oxygen
N	All nitrogen
S	All sulfur
>C=C<	Double bonds
>C=O	Carbonyl group
R—CN	Nitriles
—CEC—	Triple bonds
S=O	Sulfoxyl
—N=N—	Azo compounds
<O>	Benzene ring
<O!O>	Napthlene ring
F	All fluorine
CL	All chlorine
BR	All bromine
I	All iodine
R—OH	Alcohols
R—COOR	Esters
R—COOH	Acids
R—O—R	Ethers
<6>	Six-membered rings, (aliphatic)
<5>	Five-membered rings (aliphatic)
<0> —OH	Phenols

CRITICAL VOLUME ESTIMATER

NAME OF COMPOUND =? **PROPANE**

DO YOU WANT TO SEE THE LIST OF SYMBOLS (Y/N)? **N**

ENTER SYMBOLS FOR STRUCTURAL CONTRIBUTIONS

ENTER 'END' TO STOP

PRESS ENTER TO CONTINUE?

SYMBOL =? **C**

NO. OF UNITS PER MOLECULE =? **3**

SYMBOL =? **H**

NO. OF UNITS PER MOLECULE =? **8**

SYMBOL =? **END**

COMPOUND:PROPANE

SYMBOL	NO. PER MOLE	CONTRIBUTION
C	3	23
H	8	17

ESTIMATED CRITICAL VOLUME (ML/MOLE)=205

READY

```
    LISTING FOR BASIC PROGRAM "VC"

 10 DIM A1$(25),V(25),AA$(25),N(25),B$(25)

 20 FOR I=1 TO 24

 30 READ A1$(I),V(I)

 40 NEXT I

 50 FOR I=1 TO 25

 60 READ B$(I)

 70 NEXT I

 80 CLS

 90 PRINT"CRITICAL VOLUME ESTIMATER"

 95 INPUT"NAME OF COMPOUND=";C$

110 INPUT"DO YOU WANT TO SEE THE LIST OF SYMBOLS(Y/N)";A$

120 IF A$="Y" THEN GOSUB 1000

130 CLS

140 N=1

150 PRINT"ENTER SYMBOLS FOR STRUCTURAL CONTRIBUTIONS"

160 PRINT"ENTER 'END' TO STOP"

170 INPUT"PRESS ENTER TO CONTINUE";Z$

180 CLS

190 FLAG=1

200 INPUT"SYMBOL=";AA$(N)
```

```
210 IF AA$(N)="END" THEN 320

240 FOR I=1 TO 25

250 IF AA$(N)=A1$(I) THEN FLAG=0

260 NEXT I

270 IF FLAG=0 THEN 295

280 PRINT"SYMBOL NOT FOUND......TRY AGAIN"

285 INPUT "PRESS 'ENTER' TO CONTINUE";A$

290 GOTO 180

295 INPUT"NO. OF UNITS PER MOLECULE=";N(N)

300 N=N+1

310 GOTO 180

320 REM-BEGIN CALCULATIONS

325 VC=0

330 FOR I=1 TO N-1

340 FOR J=1 TO 24

350 IF AA$(I)<>A1$(J) THEN 370

360 VC=VC+N(I)*V(J):V2(I)=V(J)

370 NEXT J

380 NEXT I

400 CLS

410 LPRINT"COMPOUND:";C$

415 LPRINT

420 LPRINT"SYMBOL","NO. PER MOLE","CONTRIBUTION"

425 LPRINT

430 FOR I=1 TO N-1

440 LPRINT AA$(I),N(I),V2(I)

450 NEXT I

460 LPRINT
```

```
470 LPRINT "ESTIMATED CRITICAL VOLUME(ML/MOLE)=";VC

480 END

1000 LPRINT"SYMBOL","CONTRIBUTION","DESCRIPTION"

1010 FOR I=1 TO 24

1020 LPRINT A1$(I),V(I),B$(I)

1030 LPRINT

1040 NEXT I

1050 RETURN

2000 DATA C,23,H,17,O,21,N,13.5,S,58,>C=C<,13.2,>C=O,18,R-CN,60.5,-CEC-
     ,32.7,S=O,18,-N=N-,60.5,<O>,-22.5,<O!O>,-22.5,F,31,CL,63.5,BR,82
     ,I,110.9

2010 DATA R-OH,.58,R-COOR,15.45,R-COOH,19.71,R-O-R,.40,<6>,-42,<5>,-25,
     <O>-OH,-41.4

2020 DATA ALL CARBON,ALL HYDROGEN,ALL OXYGEN,ALL NITROGEN,ALL SULFUR,DO
     UBLE BONDS(ALL TYPES),CARBONYL GROUP,NITRILES,TRIPLE BONDS,SULFO
     XYL,AZO COMPOUNDS

2030 DATA BENZENE RING,NAPTHLENE RING,ALL FLUORINE,ALL CHLORINE,ALL BRO
     MINE,ALL IODINE,ALCOHOLS,ESTERS,ACIDS,ETHERS,SIX-MEMBERED RINGS(
     ALIPHATIC),FIVE-MEMBERED RINGS(ALIPHATIC),PHENOLS

3020 DATA END
```

VAPOR PRESSURE, HEAT OF VAPORIZATION, AND OTHER PROPERTIES

The program VAPOR is used to predict the following properties:

1. Boiling point with TB/TC known[4]
   Critical Compressibility factor[5]
   Acentric factor[6]
2. Heat of vaporization[7]
   Vapor pressure[8]

To use the program, the critical temperature, critical pressure, and critical volume must be known or estimated. When, using option 1 of VAPOR to predict the critical compressibility factor of a compound, enter 0 when prompted for the critical volume. The program will skip the boiling point calculations and print answers for the critical compressibility factor and the acentric factor.

Equations for VAPOR

$Tb = .012186\theta e^B$

$B = [(1-\theta)^{2/7} - .048)\ln(Vc) + (1-\theta)^{2/7} \ln(Pc) + 1.255)]/(1-\theta)^{2/7}$

$\theta = Tb/Tc$

$Zc = .371 - .0343 * \ln(Pc)/[2.303 * (1/\theta - 1)]$

$\omega = [-\ln(Pc) - 5.92714 + 6.09648(\theta)^{-1} + 1.28862\ln(\theta) - .16934(\theta)^6]/$
$\quad [15.2518 - 15.6875(\theta)^{-1} - 13.4721\ln(\theta) + .43577(\theta)^6]$

$Hv = 1.987 * Tc * Tb * \ln(Pc) * (Tc - T)^{.38}/(Tc-Tb)^{1.38}$

$Tr = T/Tc$

$X1 = Pc * Vc/(82.3 * Tc)$

$X2 = 7 + (1 - 3.72 * X1)/(.26 * X1)$

$X3 = 36/Tr + 42 * \ln(Tr) - 35 - Tr^6$

$X4 = .0364 * X3 - \ln(Tr)/2.303$

$X5 = .118 * X3 - 7 * \ln(Tr)/2.303$

$X6 = \exp[2.303 * (-X5 - (X2 - 7) * X4)]$

$PV = X6 * Pc * 760$

Nomenclature for VAPOR

VARIABLE	DESCRIPTION
Tb	Normal boiling point (K)
T	Temperature (K)
Tc	Critical temperature (K)
Pc	Critical pressure (atm)
Vc	Critical volume (ml/mole)
Zc	Critical compressibility factor
θ	Tb/Tc
ω	Acentric factor
Tr	Reduced temperature T/Tc
PV	Vapor pressure (mmHg)
Hv	Heat of vaporization (cal/mole)
X1–X6	Temporary Variables

EXAMPLE

RUN

1. ESTIMATE BOILING PT. WITH TB/TC KNOWN

2. ESTIMATE VAPOR PRESSURE AND HEAT OF VAPORIZATION

   ENTER 0–2. . . . . . . ENTER 0 TO END?   **1**

COMPOUND NAME:?   **O-XYLENE**

TB/TC = ?   **.6617**

CRITICAL PRESSURE (ATM) = ?   **36.8**

CRITICAL VOLUME (ml/mole) = ?   **369**

COMPOUND:O-XYLENE

ESTIMATED BOILING PT(C) = 138.425

ESTIMATED CRITICAL TEMP(K) = 621.996

ESTIMATED CRITICAL COMPRESSIBILITY FACTOR = .265967

ESTIMATED ACENTRIC FACTOR = .303058

1. ESTIMATE BOILING PT. WITH TB/TC KNOWN

2. ENTER VAPOR PRESSURE AND HEAT OF VAPORIZATION

   ENTER 0–2. . . . . . . .ENTER 0 TO END?   **0**

PROGRAM EXECUTION TERMINATED

READY

EXAMPLE

RUN

1. ESTIMATE BOILING PT. WITH TB/TC KNOWN

2. ESTIMATE VAPOR PRESSURE AND HEAT OF VAPORIZATION

   ENTER 0–2. . . . . . .ENTER 0 TO END?   **2**

VAPOR PRESSURE AND LATENT HEAT OF VAPORIZATION ESTIMATER

COMPOUND NAME = ?   **O-XYLENE**

BOILING PT(C) = ?   **144**

CRITICAL PRESSURE (ATM) = ?   **36.8**

CRITICAL TEMP(K) = ?   **630.2**

CRITICAL VOLUME (ml/mole)=?   **369**

STARTING TEMP(C)=?   **100**

ENDING TEMP(C)=?   **140**

INCREMENT=?   **5**

VAPOR PRESSURE ESTIMATE

COMPOUND:O-XYLENE

CRITICAL PRESSURE (ATM)=36.8   CRITICAL TEMP(K)=630.2

BOILING PT(C)=144   CRITICAL VOLUME (ml/mole)=369

TEMP(C)	VAPOR PRESS. (mmHg)	HV(cal/mole)
100	187.143	9493.71
105	222.655	9423.11
110	263.489	9351.63
115	310.217	9279.25
120	363.438	9205.94
125	423.786	9131.67
130	491.927	9056.39
135	568.552	8980.08
140	654.383	8902.7

MAKE ANOTHER RUN (Y/N)?   **N**

1. ESTIMATE BOILING PT. WITH TB/TC KNOWN

2. ESTIMATE VAPOR PRESSURE AND HEAT OF VAPORIZATION

    ENTER 0–2. . . . . . . ENTER 0 TO END?   **0**

PROGRAM EXECUTION TERMINATED

READY

```
   LISTING FOR BASIC PROGRAM "VAPOR"


10 CLS

20 PRINT"1. ESTIMATE BOILING PT. WITH TB/TC KNOWN"
```

```
30 PRINT"2. ESTIMATE VAPOR PRESSURE AND HEAT OF VAPORIZATION"

40 PRINT"    ENTER 0-2.......ENTER 0 TO END";

50 INPUT C

60 IF C=0 THEN 90

70 ON C GOSUB 1000,2000

80 GOTO 10

90 PRINT"PROGRAM EXECUTION TERMINATED"

100 END

1000 REM- CALCULATE BOILING PT, CRITICAL TEMP, ACENTRIC FACTOR, CRITICA
     L COMPRESSIBILITY FACTOR

1010 CLS

1015 INPUT"COMPOUND NAME:";C$

1020 INPUT"TB/TC=";THETA

1030 INPUT"CRITICAL PRESSURE(ATM)=";PC

1040 INPUT"CRITICAL VOLUME(ML/MOLE)=";VC

1050 ZC=.371-.0343*LOG(PC)/(2.303*(1/THETA-1))

1055 IF VC=0 THEN 1100

1060 X1=(1-THETA)[.28591

1070 X2=LOG(VC)*(X1-.048)+X1*LOG(PC)+1.255

1080 X3=X2/X1

1090 TB=.012186*THETA*EXP(X3)

1095 TC=TB/THETA

1097 TB=TB-273.15

1100 REM-CALCULATE ACENTRIC FACTOR

1110 X1=-LOG(PC)-5.92714+6.09648/THETA+1.28862*LOG(THETA)-.169347*THETA
     [6

1120 X2=15.2518-15.6875/THETA-13.4721*LOG(THETA)+.43577*THETA[6

1130 W=X1/X2

1140 CLS
```

```
1150 LPRINT"COMPOUND:";C$

1160 LPRINT

1170 LPRINT"ESTIMATED BOILING PT(C)=";TB

1180 LPRINT"ESTIMATED CRITICAL TEMP(K)=";TC

1190 LPRINT"ESTIMATED CRITICAL COMPRESSIBILITY FACTOR=";ZC

1200 LPRINT"ESTIMATED ACENTRIC FACTOR=";W

1210 LPRINT:LPRINT:LPRINT

1220 RETURN

2000 CLS

2010 PRINT"VAPOR PRESSURE AND LATENT HEAT OF VAPORIZATION ESTIMATER"

2020 PRINT

2030 INPUT"COMPOUND NAME:";C$

2035 INPUT"BOILING PT(C)=";TB:TB=TB+273.15

2040 INPUT"CRITICAL PRESSURE(ATM)=";PC

2050 INPUT"CRITICAL TEMP(K)=";TC

2060 INPUT"CRITICAL VOLUME(ML/MOLE)=";VC

2070 CLS

2080 INPUT"STARTING TEMP(C)=";T1

2090 INPUT"ENDING TEMP(C)=";T2

2100 INPUT"INCREMENT=";INC

2110 CLS

2120 LPRINT"VAPOR PRESSURE ESTIMATE"

2130 LPRINT

2140 LPRINT"COMPOUND:";C$

2150 LPRINT

2160 LPRINT"CRITICAL PRESSURE(ATM)=";PC;TAB(30);"CRITICAL TEMP(K)=";TC

2170 LPRINT"BOILING PT(C)=";TB-273.15;TAB(30);"CRITICAL VOLUME(cc/mole)
     =";VC
```

```
2180 LPRINT

2190 LPRINT"TEMP(C)","VAPOR PRESS.(mmHg)","HV(cal/mole)"

2200 LPRINT

2205 T1=T1+273.15:T2=T2+273.15

2210 FOR T=T1 TO T2 STEP INC

2220 GOSUB 3000

2225 LPRINT T-273.15,PV,,HV

2230 NEXT T

2235 LPRINT:LPRINT:LPRINT:LPRINT

2240 CLS

2250 INPUT"MAKE ANOTHER RUN(Y/N)";A$

2260 IF A$="Y" THEN 2070

2270 RETURN

3000 REM-CALCULATE VAPOR PRESSURE AND LATENT HEAT OF VAPORIZATION

3010 TR=T/TC

3020 X1=PC*VC/(82.3*TC)

3030 X2=7+(1-3.72*X1)/(.26*X1)

3040 X3=36/TR+42*LOG(TR)-35-TR[6

3050 X4=.0364*X3-LOG(TR)/2.303

3060 X5=.118*X3-7*LOG(TR)/2.303

3070 X6=EXP(2.303*(-X5-(X2-7)*X4)).

3080 PV=X6*PC*760

3090 HV=1.987*TC*TB*LOG(PC)*(TC-T)[.38/(TC-TB)[1.38

3100 RETURN
```

HEAT CAPACITY OF GASES AND LIQUIDS

The heat capacity of pure gases can be estimated making use of Dobratz's[9] equation in conjunction with Meghrebeian's[10] recommendations for the characteristic stretching and bending vibrational wave numbers. Crawford and Parr[11] have correlated these frequencies with constants of the heat capacity equation of the form

$$Cp = A + BT + CT^2$$

To use the program, the structure of the compound must be known. The program prompts you for all of the required input. As you enter each type of chemical bond, the computer automatically sums the contributions.

When the program asks for the number of rotational bonds, it is asking for the number of single bonds about which internal rotation can occur. As an example consider methyl acetate. Rotation can take place about the terminal methyl groups and the C—O bond of the ester. Hence, the number of rotational bonds would be three.

Liquid heat capacities are related to gas heat capacities by a correlation proposed by Sternling and Brown.[18] The only information needed to make an estimate of the liquid heat capacity is the structure of the compound, the critical pressure, and the critical temperature, and boiling point.

The high-pressure heat capacity of gases can be determined by applying a correction factor obtained from Figure 6.1. The critical temperature and critical pressure must be known or estimated to use the high-pressure correlation. The structural symbols and their descriptions are listed in Table 6.4.

Equations for HTCAP

$$C_p^o = 4R + n_R * R/2 + \Sigma(q_i * C_{\gamma i}) + [(3n - 6 - n_R * \Sigma q_i)/\Sigma q_i)]$$
$$* \Sigma(q_i C^\delta_i)$$
$$C_p^o = A + BT + CT^2$$
$$(C_{PL} - C_p^o)/R = (.5 + 2.2\omega) * [3.67 + 11.64 * (1 - Tr)^4 + .634 *$$
$$(1 - Tr)^{-1}]$$
$$Tr = T/Tc$$

Nomenclature for HTCAP

VARIABLE	DESCRIPTION
C_p^o	Gas heat capacity at constant pressure (cal/mole-K)
C_{LP}	Liquid heat capacity (cal/mole-K)
Tc	Critical temperature (K)
Tr	Reduced Temperature (T/Tc)
T	Temperature (K)
Σq_i	Total number of chemical bonds

VARIABLE	DESCRIPTION
q_i	Number of bonds of the qth type
$C_{\gamma i}$	Bending coefficients (A, B, C)
$C_{\delta i}$	Stretching coefficients (A, B, C)

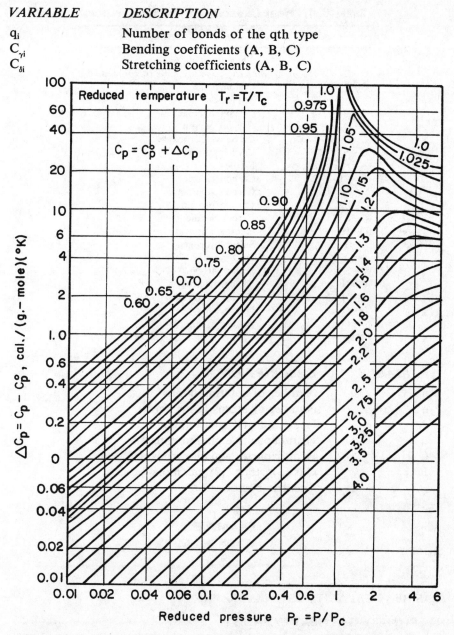

Figure 6.1   Pressure Corrections of Heat Capacities at High Pressure
Perry, R., *Chemical Engineer's Handbook,* 5th Ed., McGraw-Hill, New York, NY., 1973.

Table 6.4. Heat Capacity Structural Symbols

SYMBOL	DESCRIPTION
C—C	Single-bond aliphatic
C—C*	Single-bond aromatic or conjugated
C=C	Double bond
—CEC—	Triple bond
C—H	Carbon/hydrogen
C—O	Carbon/oxygen
C=O	Carbonyl/Carboxyl
C—N	Carbon/nitrogen
—CN	Nitrile
C—CL	Carbon/chlorine
C—F	Carbon/flourine
C—BR	Carbon/bromine
C—I	Carbon/iodine
O—H	Hydroxyl or acid
S—H	Sulfur/hydrogen
S=O	Sulfoxyl
C—S	Sulfer/carbon
N—N	Nitrogen/nitrogen
N—H	Nitrogen/hydrogen
N—O	Nitrogen/oxygen
N=O	Nitroso

VARIABLE	DESCRIPTION
n_R	Number of rotating bonds
n	Number of atoms in molecule
A	Heat capacity equation constant
B	Heat capacity equation constant
C	Heat capacity equation constant
R	Gas Constant 1.98 cal/mole-°K
ω	Acentric Factor

EXAMPLE

READY

RUN

GAS HEAT CAPACITY ESTIMATOR

1. ESTIMATE GAS HEAT CAPACITY

2. ESTIMATE LIQUID HEAT CAPACITY

ENTER CHOICE (0–2). . . . ENTER 0 TO END?   2

NAME OF COMPOUND = ?   **PROPIONITRILE**

NO. OF CHEMICAL BONDS = ?   **8**

NO. OF ROTATING BONDS = ?   **2**

NO. OF ATOMS IN MOLECULE = ?   **9**

CRITICAL PRESSURE (ATM) = ?   **41.3**

CRITICAL TEMPERATURE (K) = ?   **564**

BOILING PT (K) = ?   **351**

DO YOU WANT TO SEE THE LIST OF SYMBOLS(Y/N)?   **N**

ENTER SYMBOLS FOR STRUCTURAL CONTRIBUTIONS

ENTER 'END' TO STOP

PRESS ENTER TO CONTINUE?

SYMBOL = ?   **C—H**

NO. OF UNITS PER MOLECULE = ?   **5**

SYMBOL = ?   **C—C**

NO. OF UNITS PER MOLECULE = ?   **2**

SYMBOL = ?   **—CN**

NO. OF UNITS PER MOLECULE = ?   **1**

SYMBOL = ?   **END**

COMPOUND:PROPIONITRILE

SYMBOL	NO. PER MOLE
C—H	5
C—C	2
—CN	1

GAS HEAT CAPACITY EQUATION

CP = 7.289 + .04343790*T + − 1.4422E − 05*T**2

EXTRAPOLATE DATA (Y/N)?   **Y**

STARTING TEMP (K) = ?   **250**

ENDING TEMP (K) = ?   **350**

INCREMENT=?   **10**

TEMP (K)	LIQ. CP(CAL/MOLE)
250	26.5939
260	26.7996
270	27.0266
280	27.274
290	27.541
300	27.827
310	28.1314
320	28.4538
330	28.7941
340	29.1522
350	29.5285

CRITICAL PRESSURE(ATM)=41.3

CRITICAL TEMP=564

BOILING PT (K)=351

ACENTRIC FACTOR=.133473

MAKE ANOTHER RUN(Y/N)?   **N**

GAS HEAT CAPACITY ESTIMATOR

1. ESTIMATE GAS HEAT CAPACITY

2. ESTIMATE LIQUID HEAT CAPACITY

ENTER CHOICE (0-2). . . .ENTER 0 TO END?   **0**

PROGRAM EXECUTION TERMINATED

READY

```
   LISTING FOR BASIC PROGRAM "HTCAP"


10 R=1.987

20 DIM A1$(21),A1(21),B1(21),C1(21),A2(21),B2(21),C2(21),Q(21),AA$(21
   ),B$(21)
```

```
 30 FOR I=1 TO 21
 40 READ A1(I),B1(I),C1(I)
 50 NEXT I
 60 FOR I=1 TO 21
 70 READ A2(I),B2(I),C2(I)
 80 NEXT I
 90 FOR I=1 TO 21:READ A1$(I):NEXT I
100 FOR I=1 TO 21
110 READ B$(I)
120 NEXT I
130 CLS
140 PRINT"GAS HEAT CAPACITY ESTIMATER"
150 PRINT"1.ESTIMATE GAS HEAT CAPACITY"
160 PRINT"2.ESTIMATE LIQUID HEAT CAPACITY"
170 PRINT"       ENTER CHOICE(0-2)....ENTER 0 TO END";
180 INPUT CH
190 IF CH=0 THEN 930
200 GOSUB 950
210 INPUT"DO YOU WANT TO SEE THE LIST OF SYMBOLS(Y/N)";A$
220 IF A$="Y" THEN GOSUB 1220
230 CLS
240 N=1
250 PRINT"ENTER SYMBOLS FOR STRUCTURAL CONTRIBUTIONS"
260 PRINT"ENTER 'END' TO STOP"
270 INPUT"PRESS ENTER TO CONTINUE";Z$
280 CLS
290 FLAG=1
300 INPUT"SYMBOL=";AA$(N)
```

```
310 IF AA$(N)="END" THEN 420

320 FOR I=1 TO 21

330 IF AA$(N)=A1$(I) THEN FLAG=0

340 NEXT I

350 IF FLAG=0 THEN 390

360 PRINT"SYMBOL NOT FOUND......TRY AGAIN"

370 INPUT "PRESS 'ENTER' TO CONTINUE";A$

380 GOTO 280

390 INPUT"NO. OF UNITS PER MOLECULE=";Q(N)

400 N=N+1

410 GOTO 280

420 REM-BEGIN CALCULATIONS

430 A1=0:B1=0:C1=0:A2=0:B2=0:C2=0

440 FOR I=1 TO N-1

450 FOR J=1 TO 21

460 IF AA$(I)<>A1$(J) THEN 530

470 A1=A1+Q(I)*A1(J)

480 B1=B1+Q(I)*B1(J)

490 C1=C1+Q(I)*C1(J)

500 A2=A2+Q(I)*A2(J)

510 B2=B2+Q(I)*B2(J)

520 C2=C2+Q(I)*C2(J)

530 NEXT J

540 NEXT I

550 A1=A1+4*R+NR*R/2

560 X1=(3*NA-6-NR-Q)/Q

570 A=A1+A2*X1

580 B=(B1+B2*X1)/1.0E3
```

```
590 C=(C1+C2*X1)/1.0E6

600 LPRINT"COMPOUND:";C$

610 LPRINT:LPRINT

620 LPRINT"SYMBOL","NO. PER MOLE"

630 LPRINT

640 FOR I=1 TO N-1

650 LPRINT AA$(I),Q(I)

660 NEXT I

670 LPRINT

680 LPRINT"GAS HEAT CAPACITY EQUATION"

690 GOSUB 1060

700 CLS

710 INPUT"EXTRAPOLATE DATA(Y/N)";A$

720 IF A$="N" THEN 130

730 CLS

740 INPUT"STARTING TEMP(K)=";T1

750 INPUT"ENDING TEMP(K)=";T2

760 INPUT"INCREMENT=";INC

770 CLS:LPRINT:LPRINT

780 GOSUB 1080

790 FOR T=T1 TO T2 STEP INC

800 CP=A+B*T+C*T[2

810 IF CH=2 THEN GOSUB 1130

820 GOSUB 1170

830 NEXT T

840 LPRINT:LPRINT:LPRINT

850 IF CH=1 THEN 900

860 LPRINT"CRITICAL PRESSURE=";PC
```

```
870 LPRINT"CRITICAL TEMP(K)=";TC

880 LPRINT"BOILING PT(K)=";TB

890 LPRINT"ACENTRIC FACTOR=";W

900 INPUT"MAKE ANOTHER RUN(Y/N)";A$

910 IF A$="Y" THEN 700

920 GOTO 130

930 PRINT"PROGRAM EXECUTION TERMINATED"

940 END

950 CLS

960 INPUT"NAME OF COMPOUND=";C$

970 INPUT"NO. OF CHEMICAL BONDS=";Q

980 INPUT"NO. OF ROTATING BONDS=";NR

990 INPUT"NO. OF ATOMS IN MOLECULE=";NA

1000 IF CH=1 THEN 1050

1010 INPUT"CRITICAL PRESSURE(ATM)=";PC

1020 INPUT"CRITICAL TEMPERATURE(K)=";TC

1030 INPUT"BOILING PT(K)=";TB

1040 GOSUB 1280

1050 RETURN

1060 LPRINT"CP=";A;"+";B;"*T+";C;"*T**2"

1070 RETURN

1080 IF CH=2 THEN 1110

1090 LPRINT"TEMP(K)","CP(CAL/MOLE)","CV(CAL/MOLE)"

1100 RETURN

1110 LPRINT"TEMP(K)","LIQ. CP(CAL/MOLE)"

1120 RETURN

1130 REM-CALCULATE LIQUID HEAT CAPACITY

1140 TR=T/TC
```

```
1150 CLP=R*(.5+2.2*W)*(3.67+11.64*(1-TR)[4+.634/(1-TR))+CP

1160 RETURN

1170 IF CH=1 THEN 1200

1180 LPRINT T,CLP

1190 RETURN

1200 LPRINT T,CP,CP-R

1210 RETURN

1220 LPRINT"SYMBOL","DESCRIPTION"

1230 FOR I=1 TO 21

1240 LPRINT A1$(I),B$(I)

1250 LPRINT

1260 NEXT I

1270 RETURN

1280 REM-CALC ACENTRIC FACTOR

1290 TH=TB/TC

1300 X=-LOG(PC)-5.92714+6.09648/TH+1.28862*LOG(TH)-.16347*TH[6

1310 Y=15.2518-15.6875/TH-13.4721*LOG(TH)+.43577*TH[6

1320 W=X/Y'PERRY,R.,CHILTON,C.,CHEMICAL ENGINEERS HANDBOOK

1330 RETURN'5TH ED.,MCGRAW HILL,1973,PP.3-237,TABLE 3-316

1340 DATA -.339,3.564,-1.449

1350 DATA -.836,3.288,-1.087

1360 DATA -.740,3.730,-1.404

1370 DATA -.606,1.861,-.306

1380 DATA -.139,.168,.447

1390 DATA -.458,3.722,-1.471

1400 DATA -.778,2.721,-.759

1410 DATA -.501,3.695,-1.471

1420 DATA -.525,1.528,-.141
```

```
1430 DATA .343,2.707,-1.150

1440 DATA -.579,3.471,-1.471

1450 DATA.471,2.519,-1.076

1460 DATA .740,2.106,-.908

1470 DATA 0.000,-.240,.560

1480 DATA -.331,.805,.192

1490 DATA -.772,3.685,-1.365

1500 DATA .219,2.884,-1.218

1510 DATA -.501,3.695,-1.471

1520 DATA -.040,-.120,.530

1530 DATA -.785,3.668,-1.347

1540 DATA -.835,3.347,-1.125

1550 DATA .343,2.707,-1.150

1560 DATA .503,2.472,-1.058

1570 DATA -.339,3.564,-1.449

1580 DATA 1.268,1.244,-.544

1590 DATA -.579,3.741,-1.471

1600 DATA -.665,3.757,-1.449

1610 DATA -.034,3.220,-1.341

1620 DATA 1.016,1.663,-.723

1630 DATA 1.665,.566,-.249

1640 DATA 1.613,.656,-.289

1650 DATA -.740,3.730,-1.404

1660 DATA -.415,3.360,-1.462

1670 DATA -.275,3.498,-1.431

1680 DATA -.819,3.563,-1.267

1690 DATA -.230,3.450,-1.416
```

```
1700 DATA .774,2.031,-.886

1710 DATA 1.558,.750,-.330

1720 DATA -.320,3.540,-1.445

1730 DATA -.740,3.730,-1.404

1740 DATA .311,2.754,-1.168

1750 DATA .343,2.707,-1.150

1760 DATA C-C,C-C*,C=C,-CEC-,C-H,C-O,)C=O,C-N,-CN,C-CL,C-F,C-BR,C-I,O-H
     ,S-H,S=O,C-S,N-N,N-H,N-O,N=O

1770 DATA SINGLE BOND ALIPHATIC,SINGLE BOND AROMATIC OR CONJUGATED,DOUB
     LE BOND,TRIPLE BOND,CARBON/HYDROGEN,CARBON/OXYGEN,CARBON/YL/CARB
     OXYL,CARBON/NITROGEN,NITRILE,CARBON/CHLORINE,CARBON FLUORINE

1780 DATA CARBON/BROMINE,CARBON/IODINE,HYDROXYL OR ACID,SULFUR/HYDROGEN
     ,SULFOXYL,SULFUR/CARBON,NITROGEN/NITROGEN,NITROGEN/HYDROGEN,NITR
     OGEN/OXYGEN,NITROSO
```

VISCOSITY AND THERMAL CONDUCTIVITY OF GASES

Gas Viscosity

Gas viscosity can be predicted accurately using the correlations proposed by Yoon and Thodos.[12] The correlations require that the critical temperature, critical pressure, molecular weight, and boiling point be known or estimated. Three separate correlations are programmed to predict viscosities of nonpolar gases, hydrogen bonded gases, and polar gases.

Thermal Conductivity

Thermal conductivities of gases can be estimated using modifications of Svehla's correlation.[13] The program requires that the critical temperature, critical pressure, boiling point, and coefficients of the gas heat capacity equation be known or estimated. The program "HTCAP" can be used to obtain the coefficients of the heat capacity equation. The program then calculates the heat capacity at constant volume and the viscosity at the same temperature. These two values are used to calculate the gas thermal conductivity using the modified Svehla correlations. The estimated values are averaged and then printed.

The predicted values of viscosity and thermal conductivity are for low pressures only. Figures 6.2 and 6.3 are charts that allow you to extrapolate

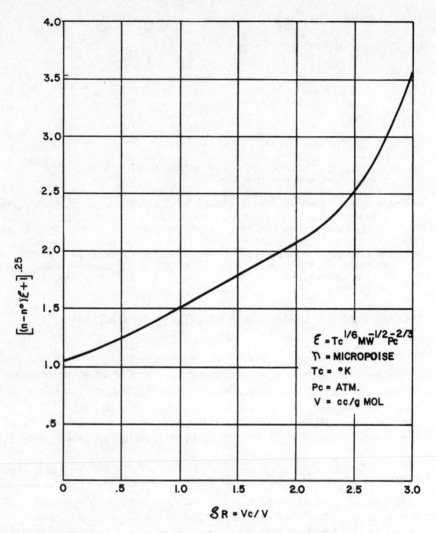

Figure 6.2   Viscosity Corrections for Gases at High Pressure

low-pressure estimates to higher pressures. To use the charts, the critical volume must be known or estimated.

Equations for GASVISC

$$\eta\xi = 4.610TR^{.618} - 2.04e^{-.449TR} + 1.94e^{-4.058TR} + .10 \text{ (nonpolar gases)}$$
$$\eta\xi = (0.755TR-0.055)Zc^{-5/4} \text{ (polar gas, hydrogen bonding)}$$

$\eta\xi = (1.90TR-.29)^{4/5}Zc^{-2/3}$ (polar gases, nonhydrogen bonding)
$\xi = Tc^{1/6}MW^{-1/2}Pc^{-2/3}$
$TR = T/Tc$
$\lambda_1 = (CV + 4.47)\,(\eta/MW)/1E6$

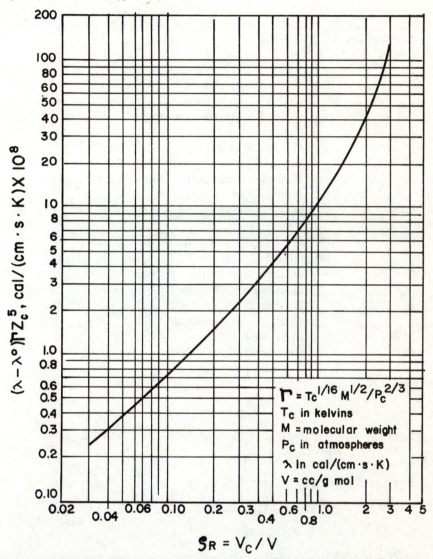

$$\S_R = V_C / V$$

Figure 6.3 Thermal Conductivity Corrections for Gases at High Pressure
Perry, R., Green, D., *Chemical Engineer's Handbook,* 6th Ed., McGraw-Hill,
New York, NY., 1984.

$$\lambda_2 = (1.32 * CV + 3.52) * (\eta/MW)/1E6$$
$$\lambda_0 = (\lambda_1 + \lambda_2) * 241.9/2$$
$$CP = A + BT + CT^2$$
$$CV = CP-R$$

Nomenclature for GASVISC

VARIABLE	DESCRIPTION
T	Temperature (K)
Tc	Critical Temperature (K)
Zc	Critical compressibility factor
ξ	Viscosity constant
η	Gas viscosity (μP)
λ_1	Thermal condition (cal/cm-s-°K)
λ_2	Thermal condition (cal/cm-s-°K)
λ_0	Thermal condition (Btu/ft-hr-°F)
CP	Heat capacity at constant pressure (cal/mole-°K)
CV	Heat capacity at constant volume (cal/mole-°K)
R	Gas constant (1.987 cal/mole-°K)
A	Heat capacity constant
B	Heat capacity constant
C	Heat capacity constant
MW	Molecular weight
Pc	Critical pressure (ATM)
TR	Reduced temperature

EXAMPLE

READY

RUN

1. ESTIMATE GAS VISCOSITY

2. ESTIMATE GAS THERMAL CONDUCTIVITY

ENTER 0-2. . . . . . .ENTER 0 TO END? **1**

COMPOUND NAME:? **PROPANE**

MOLECULAR WT = ? **44**

CRITICAL PRESSURE (ATM) = ? **41.9**

CRITICAL TEMP (K) = ?   **369.8**

NORMAL BOILING PT (K) = ?   **231.1**

WHAT TYPE OF COMPOUND?

1. NONPOLAR

2. POLAR-HYDROGEN BONDING

3. POLAR-NONHYDROGEN BONDING

COMPOUND TYPE = ?   **1**

STARTING TEMP (C) = ?   **0**

ENDING TEMP (C) = ?   **100**

INCREMENT = ?   **10**

COMPOUND:PROPANE   MOLECULAR WT = 44

CRITICAL PRESSURE (ATM) = 41.9   CRITICAL TEMP

CRITICAL COMPRESSIBILITY FACTOR = .277837   (K) = 369.1

NORMAL BOILING PT (K) = 231.1   EPSILON = .0334829

TEMP (C)	VISCOSITY (uP)
0	76.3412
10	79.1326
20	81.9146
30	84.6855
40	82.444
50	90.1891
60	92.9196
70	95.6348
80	98.3339
90	101.016
100	103.682

MAKE ANOTHER RUN(Y/N)?   **N**

1. ESTIMATE GAS VISCOSITY

2. ESTIMATE GAS THERMAL CONDUCTIVITY

     ENTER 0–2. . . . . . .ENTER 0 TO END?   **0**

PROGRAM EXECUTION TERMINATED

READY

EXAMPLE

RUN

1. ESTIMATE GAS VISCOSITY

2. ESTIMATE GAS THERMAL CONDUCTIVITY

     ENTER 0–2. . . . . . .ENTER 0 TO END?   **2**

COMPOUND NAME:?   **PROPANE**

MOLECULAR WT = ?   **44**

CRITICAL PRESSURE (ATM) = ?   **41.9**

CRITICAL TEMP (K) = ?   **369.8**

NORMAL BOILING PT (K) = ?   **231.1**

WHAT TYPE OF COMPOUND?

1. NONPOLAR

2. POLAR-HYDROGEN BONDING

3. POLAR-NONHYDROGEN BONDING

COMPOUND TYPE = ?   **1**

STARTING TEMP (C) = ?   **0**

ENDING TEMP (C) = ?   **100**

INCREMENT = ?   **10**

ENTER HEAT CAPACITY CONSTANTS. . . .USE FORM CP = A + B * T
+ C * T**2

A = ?   **2.226**

B = ?   **.061485**

C = ?   **-2.0424E-5**

COMPOUND: PROPANE   MOLECULAR WT = 44

CRITICAL PRESSURE (ATM)=41.9   CRITICAL TEMP

(K)=369.8

CRITICAL COMPRESSIBILITY FACTOR=.278307

NORMAL BOILING PT (K)=231.1   EPSILON=.0334829

$CP=2.226 + .061485*T + -2.0424E-05*T**2$

TEMP (C)	CP	CV	THERMAL CONDUCTIVITY BTU/HR-FT-DEG F
0	17.4968	15.5168	.0101951
10	17.998	16.018	.0108209
20	18.4951	16.5151	.011461
30	18.9882	17.0082	.0121149
40	19.4772	17.4972	.0127823
50	19.9621	17.9821	.0134624
60	20.4429	18.4229	.0141549
70	20.9196	18.9396	.0148593
80	21.3922	19.4122	.0155751
90	21.8608	19.8808	.0163018
100	22.3253	20.3453	.017039

MAKE ANOTHER RUN (Y/N)?   N

1. ESTIMATE GAS VISCOSITY

2. ESTIMATE GAS THERMAL CONDUCTIVITY

    ENTER 0-2. . . . . . .ENTER 0 TO END?   0

PROGRAM EXECUTION TERMINATED

READY

    LISTING FOR BASIC PROGRAM "GASVISC"

10 CLS

20 PRINT"1. ESTIMATE GAS VISCOSITY"

```
  30 PRINT"2. ESTIMATE GAS THERMAL CONDUCTIVITY"
  40 PRINT"     ENTER 0-2.......ENTER 0 TO END";
  50 INPUT C
  60 IF C=0 THEN 90
  70 ON C GOSUB 1000,2000
  80 GOTO 10
  90 PRINT"PROGRAM EXECUTION TERMINATED"
 100 END
1000 REM-CALCULATE GAS VISCOSITY
1010 GOSUB 5000
1020 GOSUB 6000
1030 GOSUB 7000
1040 LPRINT
1050 LPRINT"TEMP(C)","VISCOSITY(uP)"
1060 LPRINT
1070 T1=T1+273.15:T2=T2+273.15
1080 FOR T=T1 TO T2 STEP INC
1090 GOSUB 3000
1100 LPRINT T-273.15,VG
1110 NEXT T
1115 LPRINT:LPRINT:LPRINT:LPRINT
1120 INPUT"MAKE ANOTHER RUN(Y/N)";A$
1130 IF A$="Y" THEN 1020
1140 RETURN
2000 REM-CALCULATE THERMAL CONDUCTIVITY
2010 GOSUB 5000
2020 GOSUB 6000
2025 GOSUB 8000
```

```
2030 GOSUB 7000

2035 LPRINT"CP=";A;"+";B;"*T+";C;"*T**2"

2040 LPRINT

2050 LPRINT"TEMP(C)","CP","CV","THERMAL CONDUCTIVITY"

2060 LPRINT,,,"BTU/HR-FT-DEG F"

2070 LPRINT

2080 T1=T1+273.15:T2=T2+273.15

2090 FOR T=T1 TO T2 STEP INC

2100 GOSUB 3000

2110 GOSUB 4000

2120 LPRINT T-273.15,CP,CV,LAMBDA

2130 NEXT T

2135 LPRINT:LPRINT:LPRINT:LPRINT

2140 INPUT"MAKE ANOTHER RUN(Y/N)";A$

2150 IF A$="Y" THEN 2020

2160 RETURN

3000 REM-CALCULATE VISCOSITY

3005 TR=T/TC

3020 IF TYPE=2 THEN 3070

3030 IF TYPE=3 THEN 3090

3040 X1=4.610*TR[.618-2.04*EXP(-.449*TR)+1.94*EXP(-4.05*TR)+.10

3050 VG=X1/E

3060 RETURN

3070 VG=(.755*TR-.055)*ZC[-1.250/E

3080 RETURN

3090 VG=(1.90*TR-.29)*ZC[-.667/E

3100 RETURN
```

```
4000 REM-CALCULATE THERMAL CONDUCTIVITY

4010 CP=A+B*T+C*T[2

4020 CV=CP-R

4025 V1=VG/1E6

4030 L1=(CV+4.47)*(V1/MW)

4040 L2=(1.32*CV+3.52)*V1/MW

4050 LAMDA=(L1+L2)*241.9/2

4060 RETURN

5000 REM-INPUT DATA

5010 CLS

5020 INPUT"COMPOUND NAME:";C$

5030 INPUT"MOLECULAR WT=";MW

5040 INPUT"CRITICAL PRESSURE(ATM)=";PC

5050 INPUT"CRITICAL TEMP(K)=";TC

5055 INPUT"NORMAL BOILING PT(K)=";TB

5060 PRINT

5070 PRINT"WHAT TYPE OF COMPOUND?"

5080 PRINT"1. NONPOLAR"

5090 PRINT"2. POLAR-HYDROGEN BONDING"

5100 PRINT"3. POLAR-NONHYDROGEN BONDING"

5110 PRINT

5120 INPUT"COMPOUND TYPE=";TYPE

5130 CLS

5135 E=TC[.1667/(SQR(MW)*PC[.6667)

5140 RETURN

6000 REM-INPUT DATA

6010 CLS

6020 INPUT"STARTING TEMP(C)=";T1
```

```
6030 INPUT"ENDING TEMP=";T2

6040 INPUT"INCREMENT=";INC

6050 CLS

6060 RETURN

7000 LPRINT"COMPOUND:";C$,"MOLECULAR WT=";MW

7010 LPRINT"CRITICAL PRESSURE(ATM)=";PC,"CRITICAL TEMP(K)=";TC

7015 ZC=.371-.0343*LOG(PC)/(2.303*(TC/TB-1))

7020 LPRINT"CRITICAL COMPRESSIBILITY FACTOR=";ZC

7025 LPRINT"NORMAL BOILING PT(K)=";TB,"EPSILON=";E

7030 RETURN

8000 CLS

8010 PRINT"ENTER HEAT CAPACITY CONSTANTS....USE FORM CP=A+B*T+C*T**2"

8020 PRINT

8030 INPUT"A=";A

8040 INPUT"B=";B

8050 INPUT"C=";C

8060 CLS

8070 RETURN
```

LIQUID VISCOSITY AND DENSITY

Liquid Viscosity

The physical chemistry governing the effects that molecular structure has on liquid viscosity is not well understood. Consequently, there is no sound theoretical basis on which predictions can be made. There are numerous empirical structural correlations in the literature, but none of them is particularly accurate.

The program LIQVISC is a computerized version of the modified Orrick and Ebar[14] correlation. Orrick's original correlation was not able to make predictions for amines and sulfur compounds. The correlation method used by LIQVISC has the capability of making viscosity predictions for amines. There is no method available for making predictions for sulfur compounds.

The program requires that the molecular weight, boiling point, critical temperature, critical pressure, and structure be known. The program sums the structural contributions and then prints the predicted viscosity at the desired temperature. The symbols for the structural contributions are shown in Table 6.5. We must emphasize that when viscosity data are needed to size critical pieces of equipment, you have to get experimental values.

Liquid Density

Liquid density of saturated liquids can be estimated accurately using the relationship proposed by Goyal et al.[15] The correlation requires that the critical pressure, critical temperature, and boiling point be known. The critical compressibility factor is estimated using Edmister's[5] equation.

Equations for LIQVISC

$$\mu_L = MW\rho_L Exp\ (A + B/T)$$
$$A = \Sigma Ai - (6.95 + .21 * NC)$$

Table 6.5 Structural Symbols for Liquid Viscosity

SYMBOL	DESCRIPTION
C—R3	Tertiary carbon
C—R4	Quaternary carbon
>C=C<	Double bonds (olefins)
<5>	Five-membered rings (sat'd)
<6>	Six-membered rings (sat'd)
<O>	Benzene ring
<O>−O	Orthosubstitution
<O>−M	Metasubstitution
<O>−P	Parasubstitution
C—CL	Chloride
C—BR	Bromide
C—I	Iodine
R—OH	Alcohols and phenols
R—COOR	Esters
R—O—R	Ethers
>C=O	Ketones and aldehydes
R—COOH	Carboxyls
R—NH2	Primary amines
N<R2	Secondary amines
R—N<R2	Tertiary amines
<O>−NH2	Aniline and derivatives

$B = \Sigma Bi + 275 + 99 * NC$

$\rho_L = PcMW/Tc * [.0653/Zc^{.773} - .09*T/Tc]$

Nomenclature for LIQVISC

VARIABLE	DESCRIPTION
μ_L	Liquid viscosity (cp)
ρ_L	Liquid density (gram/ml) at 20°C
MW	Molecular weight
Pc	Critical pressure (ATM)
Tc	Critical Temperature (K)
T	Temperature (K)
Zc	Critical compressibility factor
A	Viscosity constant
B	Viscosity constant
Ai	Structural constant
Bi	Structural constant
NC	Number of carbon atoms

EXAMPLE

RUN

LIQUID VISCOSITY, DENSITY ESTIMATOR

1. ESTIMATE LIQUID VISCOSITY

2. ESTIMATE LIQUID DENSITY

ENTER CHOICE (0–2), ENTER 0 TO END?   1

COMPOUND:?   **BUTANOL**

MOLECULAR WEIGHT=?   **74**

NO. OF CARBON ATOMS=?   **4**

NORMAL BOILING PT(C)=?   **117**

CRITICAL PRESSURE(ATM)=?   **46.2**

CRITICAL TEMPERATURE(K)=?   **561**

DO YOU WANT TO SEE THE LIST OF SYMBOLS (Y/N)?   **N**

ENTER SYMBOLS FOR STRUCTURAL CONTRIBUTIONS

ENTER 'END' TO STOP

PRESS ENTER TO CONTINUE?

SYMBOL =? **R—OH**

NO. OF UNITS PER MOLECULE =? **1**

SYMBOL =? **END**

COMPOUND:BUTANOL

SYMBOL	NO. PER MOLE	A	B
R—OH	1	−3	1600

STARTING TEMP(C) =? **100**

ENDING TEMP(C) =? **120**

INCREMENT =? **2**

MOLECULAR WEIGHT = 74   NORMAL BP(C) = 117

CRITICAL PRESSURE(ATM) = 46.2   CRITICAL TEMP(K) = 561

CRITICAL COMPRESSIBILITY FACTOR = .240637

A = −10.790   B = 2271

TEMP(C)	VISCOSITY(CP)
100	.6510293
102	.590809
104	.572144
106	.554257
108	.537108
110	.52066
112	.504879
114	.489732
116	.475188
118	.461218
120	.447795

MAKE ANOTHER RUN(Y/N)? **N**

LIQUID VISCOSITY, DENSITY ESTIMATOR

1. ESTIMATE LIQUID VISCOSITY

2. ESTIMATE LIQUID DENSITY

ENTER CHOICE (0-2), ENTER 0 TO END? **0**

PROGRAM EXECUTION TERMINATED

READY

EXAMPLE

RUN

LIQUID VISCOSITY, DENSITY ESTIMATOR

1. ESTIMATE LIQUID VISCOSITY

2. ESTIMATE LIQUID DENSITY

ENTER CHOICE (0-2), ENTER 0 TO END? **2**

COMPOUND:? **BUTANOL**

MOLECULAR WEIGHT = ? **74**

NORMAL BOILING PT(C) = ? **117**

CRITICAL PRESSURE(ATM) = ? **46.2**

CRITICAL TEMPERATURE(K) = ? **561**

STARTING TEMP(C) = ? **100**

ENDING TEMP(C) = ? **120**

INCREMENT = ? **2**

MOLECULAR WEIGHT = 74   NORMAL BP(C) = 117

CRITICAL PRESSURE(ATM) = 46.2   CRITICAL

CRITICAL COMPRESSIBILITY FACTOR = .240637   TEMP(K) = 561

TEMP(C)	DENSITY(G/ML)
100	.832008
102	.830053
104	.828098

TEMP(C)	DENSITY (G/ML)
106	.826142
108	.824187
110	.822232
112	.820276
114	.818321
116	.816366
118	.81441
120	.812455

MAKE ANOTHER RUN(Y/N)?   N

LIQUID VISCOSITY, DENSITY ESTIMATOR

1. ESTIMATE LIQUID VISCOSITY

2. ESTIMATE LIQUID DENSITY

ENTER CHOICE (0–2), ENTER 0 TO END?   0

PROGRAM EXECUTION TERMINATED

READY

LISTING FOR BASIC PROGRAM "LIQVISC"

```
10 DIM A1$(21),A(21),B(21),AA$(21),B$(21),N(21),A1(21),B1(21)

20 FOR I=1 TO 21

30 READ A1$(I),A(I),B(I)

40 NEXT I

50 FOR I=1 TO 21

60 READ B$(I)

70 NEXT I

80 CLS

90 GOSUB 830
```

```
100 IF CH=2 THEN 520

110 INPUT"DO YOU WANT TO SEE THE LIST OF SYMBOLS(Y/N)";A$

120 IF A$="Y" THEN GOSUB 1060

130 CLS

140 N=1

150 PRINT"ENTER SYMBOLS FOR STRUCTURAL CONTRIBUTIONS"

160 PRINT"ENTER 'END' TO STOP"

170 INPUT"PRESS ENTER TO CONTINUE";Z$

180 CLS

190 FLAG=1

200 INPUT"SYMBOL=";AA$(N)

210 IF AA$(N)="END" THEN 320

220 FOR I=1 TO 21

230 IF AA$(N)=A1$(I) THEN FLAG=0

240 NEXT I

250 IF FLAG=0 THEN 290

260 PRINT"SYMBOL NOT FOUND......TRY AGAIN"

270 INPUT "PRESS 'ENTER' TO CONTINUE";A$

280 GOTO 180

290 INPUT"NO. OF UNITS PER MOLECULE=";N(N)

300 N=N+1

310 GOTO 180

320 REM-BEGIN CALCULATIONS

330 A=0:B=0

340 FOR I=1 TO N-1

350 FOR J=1 TO 21

360 IF AA$(I)<>A1$(J) THEN 390

370 A=A+A(J)*N(I):A1(I)=A(J)
```

```
380 B=B+B(J)*N(I):B1(I)=B(J)

390 NEXT J

400 NEXT I

410 A=A-(6.95+.21*NC)

420 B=B+275+99*NC

430 CLS

440 LPRINT"COMPOUND:";C$

450 LPRINT

460 LPRINT"SYMBOL","NO. PER MOLE","A","B"

470 LPRINT

480 FOR I=1 TO N-1

490 LPRINT AA$(I),N(I),A1(I),B1(I)

500 NEXT I

510 LPRINT

520 THETA=TB/TC

530 ZC=.371-.0343*LOG(PC)/(2.303*(1/THETA-1))

540 GOSUB 1000

550 LPRINT

560 LPRINT"MOLECULAR WEIGHT=";MW;TAB(30);"NORMAL BP(C)=";TB-273.15

570 LPRINT"CRITICAL PRESSURE(ATM)=";PC;TAB(30);"CRITICAL TEMP(K)=";TC

580 LPRINT"CRITICAL COMPRESSIBILITY FACTOR=";ZC

585 LPRINT"A=";A;TAB(30);"B=";B

590 LPRINT

600 IF CH=2 THEN 720

610 LPRINT"TEMP(C)","VISCOSITY(CP)"

620 LPRINT

630 T=293:IF TB<293 THEN T=TB

640 GOSUB 1130
```

```
650 P1=PL

660 FOR T=T1 TO T2 STEP INC

680 CP=MW*P1*EXP(A+B/T) 'CALC. VISCOSITY

690 LPRINT T-273.15,CP

700 NEXT T

710 GOTO 770

720 LPRINT"TEMP(C)","DENSITY(GRAM/ML)"

730 FOR T=T1 TO T2 STEP INC

740 GOSUB 1130

750 LPRINT T-273.15,PL

760 NEXT T

770 CLS

780 INPUT"MAKE ANOTHER RUN(Y/N)";A$

790 IF A$="Y" THEN 540

800 GOTO 90

810 PRINT"PROGRAM EXECUTION TERMINATED"

820 END

830 CLS:PRINT"LIQUID VISCOSITY, DENSITY ESTIMATER"

840 PRINT

850 PRINT"1. ESTIMATE LIQUID VISCOSITY"

860 PRINT"2. ESTIMATE LIQUID DENSITY"

870 PRINT

880 INPUT"ENTER CHOICE (0-2), ENTER 0 TO END";CH

890 IF CH=0 THEN 810

900 CLS

910 INPUT"COMPOUND:";C$

920 INPUT"MOLECULAR WEIGHT=";MW

930 IF CH=2 THEN 950
```

```
940 INPUT"NO. OF CARBON ATOMS =";NC

950 INPUT"NORMAL BOILING PT(C)=";TB

960 TB=TB+273.15

970 INPUT"CRITICAL PRESSURE(ATM)=";PC

980 INPUT"CRITICAL TEMPERATURE(K)=";TC

990 RETURN

1000 CLS

1010 INPUT"STARTING TEMP(C)=";T1

1020 INPUT"ENDING TEMP(C)=";T2

1030 INPUT"INCREMENT=";INC

1040 T1=T1+273.15:T2=T2+273.15

1050 RETURN

1060 LPRINT"SYMBOL","A","B","DESCRIPTION"

1070 FOR I=1 TO 21

1080 LPRINT A1$(I),A(I),B(I),B$(I)

1090 LPRINT

1100 NEXT I

1110 LPRINT:LPRINT:LPRINT:LPRINT

1120 RETURN

1130 PL=PC*MW/TC*((.0653/ZC[.773)-.09*T/TC)'CALC DENSITY

1140 RETURN

1150 DATA C-R3,-.15,35,C-R4,-1.20,400,)C=C<,.24,-90,(5),.10,.32,(6),-.4
     5,250,(O),0,20,(O)-O,-.12,100,(O)-M,.05,-34,(O)-P,-.01,-5,C-CL,-
     .61,220,C-BR,-1.25,365,C-I,-1.75,400,R-OH,-3.0,1600,R-COOR,-1.0,
     420,R-O-R,-.38,140,)C=O,-.50,350

1160 DATA R-COOH,-.90,770,R-NH2,.383,396,N(R2,.5252,308,R-N(R2,.2573,-1
     00.26,(O)-NH2,-2.67,1397.7

1170 DATA TERTIARY CARBON, QUATERNARY CARBON,DOUBLE BONDS,FIVE-MEMBERED
     RINGS(SAT'D),SIX-MEMBERED RINGS(SAT'D),BENZENE RING,ORTHO SUBST
     ITUTION,META SUBSTITUTION,PARA SUBSTITUTION,CHLORIDE,BROMIDE,IOD
     IDE,ALCOHOLS AND PHENOLS
```

1180 DATA ESTERS,ETHERS,KETONES AND ALDEHYDES,CARBOXYLS,PRIMARY AMINES,
   SECONDARY AMINES,TERTIARY AMINES,ANILINE AND DERIVATIVES

LIQUID THERMAL CONDUCTIVITY AND SURFACE TENSION

Liquid Thermal Conductivity

The liquid thermal conductivity of nonpolar to moderately polar liquids may be predicted with fair accuracy using the equation recommended by Reid, Prausnitz, and Sherwood.[16] The molecular weight, boiling point, and critical temperature must be known or estimated to use the program.

Liquid Surface Tension

Surface tension of nonpolar to slightly polar organics may be predicted using the correlation recommended by Reid, Prausnitz, and Sherwood.[17] The program requires that the critical temperature, critical pressure, and boiling point be known or estimated. The correlation should not be used for molecules that exhibit strong hydrogen bonding or quantum liquids (ie., H_2, He, Ne). Accuracy is generally good, with errors less than 10%.

Equations for LIQCOND

Liquid Thermal Conductivity

$$\Gamma = .6386 [3 + 20 * (1 - T/Tc)^{.667}] / \{[3 + 20 * (1 - Tb/Tr)^{.667}] * MW^{.5}\}$$

Surface Tension

$$\sigma = Pc^{.667} * Tc^{.333} * Q * (1 - T/Tc)^{1.222}$$
$$Q = .1207 * [(1 + (Tb/Tc) * \ln(Pc))/(1 - Tb/Tc)] - .281$$

Nomenclature for LIQCOND

VARIABLE	DESCRIPTION
Γ	Thermal conductivity (Btu/hr-ft-F)
σ	Surface tension (Dynes/cm)
MW	Molecular weight
T	Temperature (K)
Tc	Critical temperature (K)
Tb	Boiling point (K)

VARIABLE	*DESCRIPTION*
Pc	Critical Pressure (ATM)
Q	Temporary variable
Tr = T/Tc	Reduced temperature

EXAMPLE

RUN

1. ESTIMATE LIQUID THERMAL CONDUCTIVITY

2. ESTIMATE LIQUID SURFACE TENSION

ENTER CHOICE (0–2), ENTER 0 TO END?   **1**

COMPOUND NAME:?   **BENZENE**

MOLECULAR WT. = ?   **78**

BOILING PT.(C) = ?   **80**

CRITICAL TEMP(K) = ?   **562**

CRITICAL PRESSURE(ATM) = ?   **49**

STARTING TEMP(C) = ?   **20**

ENDING TEMP(C) = ?   **80**

INCREMENT = ?   **10**

COMPOUND:BENZENE

MOLECULAR WEIGHT = 78	BOILING PT(C) = 80
CRITICAL TEMP(K) = 562	CRITICAL PRESSURE(ATM) = 49
TEMP(C)	THERMAL COND.(BTU/Ft-HR-F)
20	.0825903
30	.0809345
40	.0792573
50	.0775575
60	.0758338
70	.0740848
80	.072309

MAKE ANOTHER RUN(Y/N)?   N

1. ESTIMATE LIQUID THERMAL CONDUCTIVITY

2. ESTIMATE LIQUID SURFACE TENSION

ENTER CHOICE (0-2), ENTER 0 TO END?   0

PROGRAM EXECUTION TERMINATED

READY

EXAMPLE

RUN

1. ESTIMATE LIQUID THERMAL CONDUCTIVITY

2. ESTIMATE LIQUID SURFACE TENSION

ENTER CHOICE (0-2), ENTER 0 TO END?   2

COMPOUND NAME:?   **BENZENE**

MOLECULAR WT. =?   **78**

BOILING PT.(C)=?   **80**

CRITICAL TEMP(K)=?   **562**

CRITICAL PRESSURE(ATM)=?   **49**

STARTING TEMP(C)=?   **20**

ENDING TEMP(C)=?   **80**

INCREMENT=?   **10**

COMPOUND:BENZENE

MOLECULAR WEIGHT=78	BOILING PT(C)=80
CRITICAL TEMP(K)=562	CRITICAL PRESSURE(ATM)=49
TEMP(C)	SURFACE TENSION (DYNES/CM)
20	28.4313
30	27.1444
40	25.8685
50	24.6039

TEMP(C)	SURFACE TENSION (DYNES/CM)
60	23.351
70	22.1103
80	20.882

MAKE ANOTHER RUN(Y/N)?   N

1. ESTIMATE LIQUID THERMAL CONDUCTIVITY

2. ESTIMATE LIQUID SURFACE TENSION

ENTER CHOICE (0-2), ENTER 0 TO END?   0

PROGRAM EXECUTION TERMINATED

READY

```
    LISTING FOR BASIC PROGRAM "LIQCOND"


10 CLS

20 PRINT"1. ESTIMATE LIQUID THERMAL CONDUCTIVITY"

30 PRINT"2. ESTIMATE LIQUID SURFACE TENSION"

40 PRINT

50 INPUT"ENTER CHOICE (0-2), ENTER 0 TO END";CH

60 IF CH=0 THEN 310

70 CLS

80 INPUT"COMPOUND NAME:";C$

90 INPUT"MOLECULAR WT.=";MW

100 INPUT"BOILING PT.(C)=";TB

110 TB=TB+273.15

120 INPUT"CRITICAL TEMP(K)=";TC

130 INPUT"CRITICAL PRESSURE(ATM)=";PC

140 GOSUB 500'GET TEMP RANGES

150 REM-BEGIN CALCULATIONS
```

```
160 LPRINT"COMPOUND:";C$

170 LPRINT

180 LPRINT"MOLECULAR WEIGHT=";MW,"BOILING PT(C)=";TB-273.15

190 LPRINT"CRITICAL TEMP(K)=";TC,"CRITICAL PRESSURE(ATM)=";PC

200 LPRINT

210 GOSUB 600'PRINT HEADERS

220 FOR T=T1 TO T2 STEP INC

230 ON CH GOSUB 1000,2000

240 LPRINT T-273.15,X

250 NEXT T

255 LPRINT:LPRINT:LPRINT

260 CLS

270 INPUT"MAKE ANOTHER RUN(Y/N)";A$

280 IF A$="Y" THEN 140

290 GOTO 10

300 PRINT"PROGRAM EXECUTION TERMINATED"

310 PRINT"PROGRAM EXECUTION TERMINATED"

320 END

500 CLS

510 INPUT"STARTING TEMP(C)=";T1

520 INPUT"ENDING TEMP(C)=";T2

530 INPUT"INCREMENT=";INC

540 T1=T1+273.15:T2=T2+273.15

550 RETURN

600 IF CH=2 THEN 630

610 LPRINT"TEMP(C)","THERMAL COND.(BTU/FT-HR-F)"

620 RETURN
```

```
 630 LPRINT"TEMP(C)","SURFACE TENSION(DYNES/CM)"

 640 RETURN

1000 REM-CALC THERMAL COND

1010 TR=T/TC:T0=TB/TC

1020 X1=3+20*(1-TR)[.667

1030 X2=3+20*(1-T0)[.667

1040 L=.00264*X1/(X2*SQR(MW))

1050 X=241.9*L

1060 RETURN

2000 REM-CALC SURFACE TENSION

2010 TR=T/TC:T0=TB/TC

2020 Q=.1207*(1+T0*LOG(PC)/(1-T0))-.281

2030 SIG=PC[.667*TC[.333*Q*(1-TR)[1.222

2040 X=SIG

2050 RETURN
```

REFERENCES

1. Gambill, W., *Chem. Eng.*, June 15, 1959, pp. 182–183.
2. Gambill, W., *Chem. Eng.,* July 13, 1959, pp. 157–160.
3. Gold, P., *Chem. Eng.,* Nov. 4, 1968, pp. 185–190.
4. Miller, C.O.M. Private communication, *Perry's Chemical Engineers Handbook,* 6th ed., McGraw-Hill, NY., NY., 1983, pp. 3–267.
5. Edmister, W., *Petrol. Refiner,* April, 1958, p. 178.
6. Lee, B., Kesler, M., *Am. Inst. Chem. Eng. J.,* 1975, 21, p. 510.
7. Perry, R., *Chemical Engineer's Handbook,* 5th ed., McGraw-Hill, NY., NY., 1973, pp. 3–239, eq. 3–54.
8. Gold P., *Chem. Eng.,* Sept. 8, 1969, p. 150.
9. Dobratz, *Ind. Eng. Chem.,* 1941, 33, p. 459.
10. Meghrebleian, *J. Am. Rocket Soc.,* 1951, 21, p. 127.
11. Crawford, Parr, *J. Chem. Phys.,* 1949, 41, p. 1037.
12. Yoon, P., Thodos, G., *AiChe J.,* 1970, 16, p. 300.
13. Svhela, R., *Estimation of Viscosity and Thermal Conductivities of Gases at High Temperature,* NASA Tech. Rep. R–132, Lewis Research Center, Cleveland, Ohio, 1962.

14. Reid, R., Praunitz, J., Sherwood, T., *The Properties of Liquids and Gases,* 3rd ed., McGraw-Hill, NY., NY., 1977, p. 437.
15. Goyal, *Hydrocarbon Proc. and Petrol.,* 1966, p. 200.
16. Reid, R., Praunitz, J., Sherwood, T., *The Properties of Liquids and Gases,* 3rd ed., McGraw-Hill, NY., NY., 1977, p. 524.
17. IBID., p. 608.
18. Perry, R., *Chemical Engineers Handbook,* 6th ed., McGraw-Hill, NY., NY., 1984, pp. 3–278, eq. 3–80.

7
POLLUTION CONTROL

POLLUTION DISPERSION

There are various steady-state atmospheric dispersion models available to the practicing engineer. STACK is based on the Texas Episodic Model (TEM)[1] developed by the Texas Air Pollution Board. The program calculates center-line ground-level concentrations of pollutants for a given stack height and weather conditions. Data for each run are stored as data statements starting at line 5000. The data are entered in the following order:

1. Pollutant emission rate (gram/sec)
2. Stack height (m)
3. Mixing height (m)
4. Wind speed (m/sec)
5. Stability class (A, B, C, DD, DN, E, F)

Atmospheric stability increases starting with class A (very unstable) and progressing to class F (moderately stable). Table 7.1 lists the Pasquill stability classes with an explanation of their ranking. A subroutine of the program will calculate effective stack heights when you want an estimate of the plume rise.

All pollutant concentrations are calculated for a 10-min. averaging period. You can select any averaging period up to 3 hr.

Equations for STACK

$$Zl = A * X^B$$
$$Yl = Cl * X^{Dl}$$

---

[1]*Users Guide-Texas Episodic Model,* Texas Air Control Board, Austin Texas, October, 1979.

Table 7.1. Pasquill Stability Classes

CLASS	STABILITY
A	very unstable (high lapse rate)
B	moderately unstable
C	slightly unstable
DD	neutral-day
DN	neutral-night
E	slightly stable
F	moderately stable (temp inversion)

$$CO = [(Q * 1E6)/(3.14 * Yl * Zl * U)] * \exp [-5 * (H/Z1)^2]$$
$$\text{If } X > 2 * XM \text{ then } CO = (Q * 1E6)/(2.506 * Yl * L * U)$$
$$A = D^2 * 3.14/4$$
$$Q_2 = [Q_1 * (T_1 + 460)/460]/34.328$$
$$Vl = Q_2/(A * 3600 * 35.28)$$
$$X3 = 1.5 + 2.609 * (T_1 - T_2)/T_1 * D$$
$$H_1 = Vl * D/U * X3$$
$$H = H + H_1$$

Nomenclature for STACK

VARIABLE	DESCRIPTION
H	Stack height (m)
Q	Emission Rate (gram/sec)
U	Wind velocity (m/sec)
L	Mixing height (m)
X	Distance downwind
XM	$XM = X$ at $Z1 = .47L$ (m)
Zl	Dispersion coefficient
Yl	Dispersion coefficient
A	Dispersion coefficient
B	Dispersion coefficient
Cl	Dispersion coefficient
Dl	Dispersion coefficient
CO	Pollutant Concentration ($\mu g/m^3$)
Al	Stack area (m^2)
Q_1	Gas flow (SCFH)
Q_2	Corrected gas flow (m^3/hr)
T_1	Stack temperature (F)
T_2	Air temperature (F)

VARIABLE	DESCRIPTION
H_1	Plume rise (m)
X3	Temperature variable
CFH	Cubic feet/hour
SCFH	Standard cubic feet/hour (60°F, 14.7 psia)
Lapse rate	Temperature gradient in atmosphere °F/1000 ft, usually negative.

EXAMPLE

LOAD"STACK"

READY

RUN

STACK DISPERSION MODEL

STACK ID (M)=? **1**

GAS FLOW(SCFH)=? **450000**

STACK TEMP(F)=? **120**

AIR TEMP(F)=? **60**

NO. OF RUNS=? **1**

AVERAGING TIME (MINUTES)=? **10**

TYPE MAX DOWNWIND DISTANCE (M), INCREMENT(M)? **2000,100**

STABILITY CLASS=A

EFF. STACK HT(M)=69.9502   MIXING HT(M)=10000

SOURCE RATE=82.5 gram/sec   WIND SPEED(M/SEC)=1.5

DISTANCE-M	PHI Y	PHI Z	CONC.-uG/CU M
100	27.5807	13.9701	.163467
200	50.5131	33.9483	1222.61
300	71.9668	57.0678	2012.15
400	92.5132	82.4969	1602.04
500	112.41	109.794	1158.53
600	131.805	161.518	749.123

DISTANCE-M	PHI Y	PHI Z	CONC.-uG/CU M
700	150.791	222.881	496.131
800	169.435	294.59	341.169
900	187.784	376.77	243.34
1000	205.876	469.53	179.203
1100	223.739	572.97	135.62
1200	241.396	687.183	105.046
1300	258.868	812.252	82.9954
1400	276.169	948.253	66.7041
1500	293.314	1095.26	54.4123
1600	310.315	1253.34	44.9661
1700	327.181	1422.56	37.5881
1800	343.921	1602.98	31.742
1900	360.543	1794.65	27.05
2000	377.055	1997.63	23.2406

AVERAGING TIME = 10 MINUTES

PROGRAM EXECTUTION TERMINATED

READY

```
   LISTING FOR BASIC PROGRAM "STACK"

10 FOR I=1 TO 7:READ A$(I):NEXT I

20 DATA A,B,C,DD,DN,E,F

30 FOR I=1 TO 7

40 READ A1(I),B1(I):NEXTI

50 FOR I=1 TO 7

60 READ A2(I),B2(I):NEXT I

70 FOR I=1 TO 7

80 READ A3(I),B3(I):NEXT I
```

```
 90 FOR I=1 TO 7

100 READ C1(I),D1(I):NEXT I

110 FOR I=1 TO 7

120 READ C2(I),D2(I):NEXT I

130 GOSUB 600

150 INPUT"NO. OF RUNS=";N

160 INPUT"AVERAGING TIME (MINUTES)=";AVG

170 INPUT"TYPE  MAX DOWNWIND DISTANCE(M), INCREMENT(M)";DW,DX

180 FOR IZ=1 TO N

190 XM=0

200 READ Q,H,L,U,A$

205 GOSUB 690

210 CLS:LPRINT"STABILITY CLASS=";A$

220 LPRINT"EFF. STACK HT(M)=";H;"    ";"MIXING HT(M)=";L

230 LPRINT

240 LPRINT"SOURCE RATE=";Q;"g/sec";"    ";"WIND SPEED(M/SEC)=";U

250 LPRINT:LPRINT

260 LPRINT"DISTANCE-M";TAB(20);"PHI Y";TAB(30);"PHI Z";TAB(40);"CONC.-
    uG/CU M"

270 LPRINT"-----------------------------------------------------------"

280 FOR X=DX TO DW STEP DX

290 GOSUB 400

300 IF XM<>0 THEN 320

310 IF Z1>=.47*L THEN XM=X

320 GOSUB 510

330 LPRINT X;TAB(20);Y1;TAB(30);Z1;TAB(40);CO*(10/AVG)[.20

340 NEXT X

350 LPRINT"AVERAGING TIME=";AVG;"MINUTES"

360 LPRINT CHR$(12)
```

```
370 NEXT I2

380 PRINT"PROGRAM EXECUTION TERMINATED"

390 END

400 FOR I=1 TO 7

410 IF A$=A$(I) THEN I1=I

420 NEXT I

430 IF X<=10000 THEN C1=C1(I1):D1=D1(I1)

440 IF X>10000 THEN C1=C2(I1):D1=D2(I1)

450 IF X<=500 THEN A=A1(I1):B=B1(I1)

460 IF X>500 AND X<=5000 THEN A=A2(I1):B=B2(I1)

470 IF X>5000 THEN A=A3(I1):B=B3(I1)

480 Z1=A*X[B

490 Y1=C1*X[D1

500 RETURN

510 REM- DO DISPERSION CALCULATIONS

520 IF XM=0 THEN 540

530 IF X>=2*XM THEN 560

540 CO=((Q*1E6)/(3.14*Y1*Z1*U))*EXP(-.5*(H/Z1)[2)

550 GOTO 570

560 GOSUB 580

570 RETURN

580 CO=(Q*1E6)/(2.506*Y1*L*U)

590 RETURN

600 REM-DO PLUME RISE CALCULATIONS

610 CLS

615 PRINT"STACK DISPERSION MODEL":PRINT

620 INPUT"STACK ID(M)=";D

630 INPUT"GAS FLOW(SCFH)=";Q1
```

```
640 INPUT"STACK TEMP(F)=";T1

650 INPUT"AIR TEMP(F)=";T2

660 T1=(T1-32)*5/9+273

670 T2=(T2-32)*5/9+273

680 A=D[2/4*3.1416

682 Q2=Q1*T1/288

684 V1=Q2/(A*3600*35.28)

685 RETURN

690 REM-CALCULATE PLUME RISE

700 X3=1.5+2.609*(T1-T2)/T1*D

710 H1=V1*D/U*X3

720 H=H+H1

730 RETURN

1000 DATA .0383,1.281,.1393,.9467,.1120,.91,.0856,.865,.0818,.8155,.109
     4,.7657,.05645,.8050

1010 DATA .0002539,2.089,.04936,1.114,.1014,.926,.2591,.6869,.2527,.634
     1,.2452,.6358,.1930,.6072

1020 DATA .0002539,2.089,.04936,1.114,.1154,.9109,.7368,.5642,1.297,.44
     21,.9204,.4805,1.505,.3662

1030 DATA .495,.873,.310,.897,.197,.908,.122,.916,.122,.916,.0934,.912,
     .0625,.911

1040 DATA .606,.851,.523,.840,.285,.867,.193,.865,.193,.865,.141,.868,.
     080,.884

2000 DATA 82.5,64,10000,1.5,A
```

CYCLONE EVALUATION AND DESIGN

CYCLONE 1 and CYCLONE 2 are useful programs for evaluating the performance of an existing cyclone or designing a new cyclone to give a specific pressure drop.

CYCLONE 1 evaluates existing cyclones when it is supplied with the operating conditions. You must supply the cyclone dimensions as defined in Figure 7.1, the gas flow rate, and the pressure, temperature, gas viscosity, gas molecular weight, and density of the dust being collected. When you

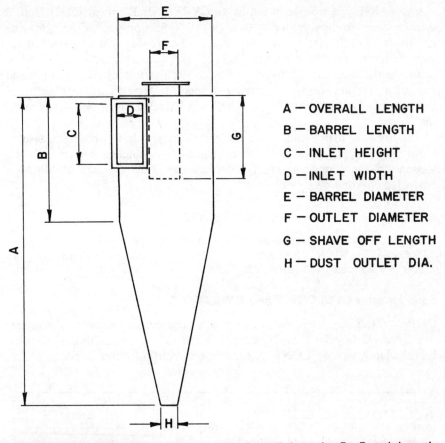

A — OVERALL LENGTH
B — BARREL LENGTH
C — INLET HEIGHT
D — INLET WIDTH
E — BARREL DIAMETER
F — OUTLET DIAMETER
G — SHAVE OFF LENGTH
H — DUST OUTLET DIA.

Figure 7.1 Cyclone Nomenclature. A—Overall Length; B—Barrel Length; C—Inlet Height; D—Inlet Width; E—Barrel Diameter; F—Outlet Diameter; G—Shave off Length; H—Dust Outlet Diameter.

have the particle size distribution of the dust being collected, the program will make an estimate of the overall collection efficiency of the cyclone. When you do not have a specific particle size distribution available, the program calculates overall efficiencies based on standardized particle size distributions. All data, except the particle size distribution, are entered through the keyboard. If you wish to use a specific distribution, the data must be entered on line 2000 as a data statement. First write the weight percent of the dust fraction (decimal), then the partical diameter in microns, for example,

2000 DATA .54,109,.06,67,.08,50

CYCLONE 2 is similar in nature to CYCLONE 1, except CYCLONE 2 is more suited to the design of new cyclones. You supply all of the operating conditions and the desired pressure drop. The program calculates all of the cyclone dimensions as shown in Figure 7.1. You may choose one of three dust classifications to get an idea of the overall efficiency of the cyclone at the given pressure drop. You may then specify a new dust classification or a new pressure drop and the program will immediately recalculate the dimensions.

The pressure drops and collection efficiencies in both programs assume a dust loading of 1 grain/ft³. If the dust loading is higher, the cyclone collection efficiency must be corrected using Fig. 7.2 from API Publication No. 931, *Cyclone Separators*.

REFERENCES

Koch, W.; Licht, W., Chem. Eng., Nov. 7, 1977, pg. 80–88
API Publication 931, *Cyclone Separators,* Chapter 11, American Petroleum Institute, Washington, D.C., 1975, Chapter 11

Equations for CYCLONE 1 and CYCLONE 2

Cyclone efficiency is dependent to a large degree on geometry. Geometry determines the cyclone configuration factor G. In CYCLONE 1, G must be calculated in CYCLONE 2, the cyclone configuration factor is 551.3,

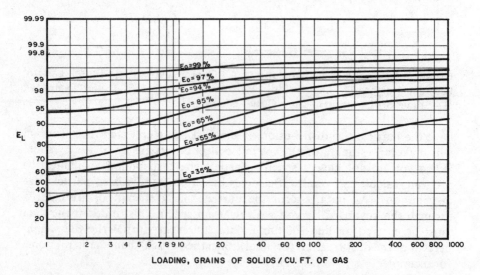

Figure 7.2 Efficiency Correction for Dust Loading
Reprinted Courtesy of the American Petroleum Institute.

the value for a Stairmind high-efficiency cyclone. For CYCLONE 1, we only list those equations used to calculate G. For CYCLONE 2, we list equations common to both programs; they are mainly the equations used to calculate efficiency after the cyclone configuration factor is known.

Equations used by CYCLONE 1 to calculate G

$$L = 2.3 * DE * DC^2/(A * B)^{.333}$$
$$VS = [\pi * (S - A/2) * (DC^2 - DE^2)]/4$$

For L > (H − S)

$$D = DE - (DC - B1) * (S + L - H1) / (H - H1)$$
$$VN = [\pi * DC^2 * (H1 - S)/4] + (\pi * DC^2/4) * [(L + S - H1)/3] * (1 + D/DC + (D/DC)^2) - (\pi * L * DE^2/4)$$
$$KC = (2 * VS + VN)/2 * DC^3$$

For L < (H − S)

$$VH = [\pi * DC^2 * (H1 - S)/4] + (\pi * DC^2/4) * ((H - H1)/3 * (1 + B1/DC + (B1/DC)^2) - (\pi * DE^2/4) * (H - S)$$
$$KC = (2 * VS + VH)/(2 * DC^3)$$
$$G = 8 * KC/((A/DC)^2 * (B/DC)^2)$$

Pressure Drop

$$DH = PF * NH * .003 * V1^2$$
$$NH = 16 * A * B/DE^2$$

Nomenclature for CYCLONE 1

VARIABLE	DESCRIPTION
DC (D)	Cyclone diameter (ft)
DE	Outlet Diameter (ft)
A	Inlet height (ft)
B	Inlet weight (ft)
B1	Dust outlet (ft)
H	Overall length (ft)
H1 h	Barrel length (ft)
S	Vortex tube (shave off) (ft)
VH	Volume below exit tube (ft³)
KC	Cyclone volume constant
G	Cyclone configuration factor
DH	Pressure drop (in H₂0)

VARIABLE	DESCRIPTION
NH	Number of inlet velocity heads
π	3.1416
PP	Solid density (lb/ft^3)
PF	Gas density (lb/ft^3)
T	Temperature (F)
P	Pressure (psia)
VS	Annular volume above exist duct (ft^3)
L	Length below exit where vortex turns (ft)
V1	Inlet velocity (ft/sec) (Defined below)

Equations for CYCLONE 2

$$DC = (1.92 * PF * Q1^2/HW)^{.25}$$
$$DE = DC/2$$
$$A = DC/2$$
$$B = DC * .20$$
$$S = DC/2$$
$$H = D * 1.5$$
$$H1 = DC * 4$$
$$B1 = DC * .375$$
$$V1 = Q1/(A * B)$$
$$V2 = Q1/(DE^2/4 * \pi)$$
$$W = [(A * 32.2 * CP * (PP - PF) * .000674)/(3 * PF^2)]^{.333}$$
$$VS = 2.055 * W * [(B/DC)/(B/DC)^{.40}] * DC^{.067} * V1^{.667}$$
$$N = 1 - [1 - (12/DC)^{.14}/2.5] * [(T + 460)/530]^{.30}$$
$$TAU = PP * DP^2/(18 * .00067 * CP)$$
$$X = 551.3 * TAU * Q1/DC^3 * (N + 1)$$
$$E = 1 - \exp(-2 * X^{.5/(N + 1)})$$
$$ET = \sum_{i=1}^{n} E * W_i$$

Nomenclature for CYCLONE 2

VARIABLE	DESCRIPTION
DC	Cyclone Barrel diameter (ft) D
DE	Outlet diameter (ft)
A	Inlet height (ft)
B	Inlet width (ft)
S	Shave-off length (ft)
H	Barrel length (ft)
H1	Overall length (ft)
B1	Dust outlet diameter (ft)

VARIABLE	*DESCRIPTION*
V1	Inlet velocity (ft/sec)
V2	Outlet velocity (ft/sec)
VS	Saltation velocity (ft/sec)
N	Vortex exponent
TAU	Relaxation time (sec)
X	Temporary variable
E	Fraction efficiency (Decimal)
ET	Overall Efficiency (Decimal)
Q1	Flow rate (ACFH)
PF	Gas density (lb/ft^3)
PP	Particle density (lb/ft^3)
CP	Gas viscosity (cp)
T	Gas temperature (°F)
ACFH	Actual Cubic Feet Per Hour
DP	Particle diameter (ft)
W_i	Weight fraction (decimal)

EXAMPLE

RUN "CYCLONE1"

CYCLONE EVALUATION

CYCLONE DIA-FT?   **2**

INLET WIDTH, INLET HEIGHT-FT?   **.5,1.5**

GAS OUTLET DIA, VORTEX TUBE LENGTH-FT?   **1,2**

BARREL LENGTH, OVERALL LENGTH-FT?   **3,6**

DUST OUTLET DIA-FT?   **.50**

GAS FLOW-SCFH, PRESS-PSIA?   **162000,14.7**

GAS TEMP-F, GAS VISCOSITY-CP?   **100,.015**

GAS MOLECULAR WT.?   **29**

SOLIDS DENSITY-LBS./CU. FT.?   **162**

TABLE OF RESULTS

CYCLONE DIMENSIONS

CYCLONE DIA-FT=2   BARREL LENGTH-FT=3

OVERALL LENGTH-FT=6   VORTEX TUBE-FT=2

TABLE OF RESULTS (*CONT.*)

GAS OUTLET DIA-FT = 1	DUST OUTLET DIA-FT = .5
INLET WIDTH-FT = .5	INLET LENGTH-FT = 1.5
GAS FLOW-SCFH = 162000	PRESS-PSIA = 14.7
GAS TEMP-F = 100	GAS VISCOSITY-CP = .015
GAS MW = 29	SOLIDS-#/FT^3 = 162

?

DO YOU HAVE PARTICLE SIZE DATA (Y/N) ?   N

DO YOU WANT AN ESTIMATE OF CYCLONE EFFICIENCY (Y/N) ?   Y

CHOOSE DUST CLASSIFICATION (1–3)

1. COARSE

2. FINE

3. SUPER FINE

?   2

DP-MICRONS	WEIGHT %	GRADE EFF	% ESCAPING
125	3	99.9979	2.11721E–03
90	7	99.9847	.015295
67	10	99.9338	.0661761
50	15	99.7778	.222216
35	10	99.2558	.744239
25	10	98.1322	1.86784
17.5	7	95.8947	4.10535
12.5	8	92.5232	7.47684
9	4	87.9585	12.0415
6.25	6	81.5418	18.4582
3.75	8	70.8346	29.1654
1.5	12	50.3108	49.6892

OVERALL EFFICIENCY = 88.9271%

TRY ANOTHER CLASSIFICATION (Y/N)?   N

VELOCITY AND PRESSURE DROP DATA

PRESSURE DROP-INCHES H20 = 12.745

INLET VELOCITY-FT/SEC = 73.0435

INLET VELOCITY/SALTATION VELOCITY = 1.77093

OUTLET VELOCITY-FT/SEC = 69.7588

READY

EXAMPLE

READY

RUN "CYCLONE 2"

CYCLONE DESIGN PROGRAM

FLOW-SCFH:?   **50000**

PRESSURE-PSIA:?   **14.7**

TEMP-F:?   **500**

GAS VISCOSITY-CP:?   **.0125**

PARTICLE DENSITY-LB/FT3:?   **162**

GAS MOLECULAR WEIGHT:?   **30**

ALLOWABLE PRESSURE DROP-IN. H20:?   **5**

DO YOU WANT AN ESTIMATE OF CYCLONE EFFICIENCY-(Y/N)?   **Y**

CHOOSE DUST CLASSIFICATION

1. COARSE

2. FINE

3. SUPER FINE

CHOOSE CLASSIFICATION (1–3)?   **2**

TABLE OF RESULTS

FLOW-SCFH = 50000   PRESS-PSIA = 14.7

TEMP-F = 500   GAS VISCOSITY-CP = .0125

MOLE WT = 30   SOLIDS DENSITY-LB/FT3 = 162

ALLOWABLE PRESS DROP (IN/H20) = 5

CALCULATED CYCLONE DIMENSIONS

INLET HEIGHT (FT) = .94792   INLET WIDTH (FT) = .379168

BARREL DIA (FT) = 1.89584   OUTLET DIA (FT) = .94792

BARREL LENGTH (FT) = 2.84376   OVERALL LENGTH (FT) = 7.58336

SHAVE OFF (FT) = .94792   DUST OUTLET DIA (FT) = .71094

INLET VEL (FT/SEC) = 80.645

INLET VELOCITY/SALTATION VELOCITY = 1.59215

OUTLET VEL (FT/SEC) = 41.0765

ESTIMATED CYCLONE EFFICIENCY-FINE

DP-MICRONS	WT-%	%-COLLECTED
125	3.0	100
90	7.0	100
67	10.0	99.9995
50	15.0	99.9948
35	10.0	99.9539
25	10.0	99.7676
17.5	7.0	99.1092
12.5	8.0	97.594
9	4.0	94.8144
6.25	6.0	89.8803
3.75	8.0	79.8131
1.5	12.0	56.8622

OVERALL EFFICIENCY = 92.1104%

DO YOU WANT TO EXAMINE ANOTHER PARTICLE DISTRIBUTION
(Y/N) ? N

CHANGE PRESSURE DROP (Y/N) ? N

PROGRAM EXECUTION TERMINATED

READY


```
   LISTING FOR BASIC PROGRAM "CYCLONE1"

 2 DIM DP(20),W(20)

 5 PI=3.1416

 7 CLS

10 PRINT "CYCLONE EVALUATION":PRINT

20 INPUT"CYCLONE DIA-FT";DC

30 INPUT"INLET WIDTH,INLET HEIGHT-FT";B,A

40 INPUT"GAS OUTLET DIA, VORTEX TUBE LENGTH-FT";DE,S

50 INPUT"BARREL LENGTH, OVERALL LENGTH-FT";H1,H

60 INPUT"DUST OUTLET DIA-FT";B1

70 CLS

80 INPUT"GAS FLOW-SCFH, PRESS-PSIA";Q,P

90 INPUT"GAS TEMP-F, GAS VISCOSITY-CP";T,CP

95 INPUT"GAS MOLECULAR WT.";MW

100 INPUT"SOLIDS DENSITY-LBS./CU. FT.";PS:CLS

160 CLS

170 L=2.3*DE*((DC*DC)/(A*B))[.333

180 VS=(PI*(S-A/2)*(DC[2-DE[2))/4

190 IF L)(H-S) THEN 240

200 D=DC-(DC-B1)*(S+L-H1)/(H-H1)

210 VN=(PI*DC[2*(H1-S)/4)+(PI*DC[2/4)*((L+S-H1)/3)*(1+D/DC+(D/DC)[2)-(
    PI*L*DE[2/4)
```

inputs

calc

```
220 KC=(2*VS+VN)/(2*DC[3)

230 GOTO 260

240 VH=(PI*DC[2*(H1-S)/4)+(PI*DC[2/4)*((H-H1)/3)*(1+B1/DC+(B1/DC)[2)-(
    PI*DE[2/4)*(H-S)

250 KC=(2*VS+VH)/(2*DC[3)

260 G=8*KC/((A/DC)[2*(B/DC)[2)

270 C1=359*((460+T)/460)*14.7/P

280 PG=MW/C1

290 W=((.0868*CP*(PS-PG))/(3*PG[2))[.333

300 Q1=Q*C1/(359*3600)

310 VI=Q1/(A*B)

320 X=(B/DC)/(B/DC)[.400

330 VS=2.055*W*X*DC[.067*VI[.667

340 X=1-(12*DC)[.14/2.5

350 Y=((T+460)/530)[.30

360 N1=1-X*Y

370 PRINT TAB(20);"TABLE OF RESULTS"

380 PRINT:PRINT TAB(20);"CYCLONE DIMENSIONS"

390 PRINT"CYCLONE DIA-FT=";DC;TAB(40);"BARREL LENGTH-FT=";H1

400 PRINT"OVERALL LENGTH-FT=";H;TAB(40);"VORTEX TUBE-FT=";S

410 PRINT"GAS OUTLET DIA-FT=";DE;TAB(40);"DUST OUTLET DIA-FT=";B1

420 PRINT"INLET WIDTH-FT=";B;TAB(40);"INLET LENGTH-FT=";A

430 PRINT

440 PRINT"GAS FLOW-SCFH=";Q;TAB(40);"PRESS-PSIA=";P

450 PRINT"GAS TEMP-F=";T;TAB(40);"GAS VISCOSITY-CP=";CP

460 PRINT"GAS MW=";MW;TAB(40);"SOLIDS-#/FT3=";PS

470 INPUT A$

480 CLS

482 GOSUB 1000
```

```
485 PRINT"DP-MICRONS","WEIGHT %","GRADE EFF","% ESCAPING"

487 W2=0

490 FOR I=1 TO N

500 DPI=DP(I)*3.25E-6

510 TAU=(DPI[2*PS)/(18*CP*6.74E-4)

520 X=(G*TAU*Q1*(N1+1)/DC[3)[(.5/(N1+1))

530 N2=1-EXP(-2*X)

540 W2=W2+W(I)*N2

550 PRINT DP(I),W(I)*100,N2*100,(W(I)-W(I)*N2)/W(I)*100

560 NEXT I

570 PRINT "OVERALL EFFICIENCY=";W2*100;"%"

580 INPUT"TRY ANOTHER CLASSIFICATION(Y/N)";A$

582 IF A$="N" THEN 590

584 GOSUB 1100

586 GOTO 485

590 CLS

600 NH=16*A*B/(DE[2)

610 DH=PG*NH*.003*VI[2

615 PRINT"VELOCITY AND PRESSURE DROP DATA"

617 PRINT"==================================="

620 PRINT"PRESSURE DROP-INCHES H20=";DH

630 PRINT"INLET VELOCITY-FT/SEC=";VI

640 PRINT"INLET VELOCITY/SALTATION VELOCITY=";VI/VS

650 PRINT"OUTLET VELOCITY-FT/SEC=";Q1/(DE[2*PI/4)

660 END

1000 REM-PARTICLE SIZE ROUTINE

1010 INPUT"DO YOU HAVE PARTICLE SIZE DATA(Y/N)";A$

1020 IF A$="N" THEN 1080

1030 INPUT"HOW MANY WEIGHT FRACTIONS";N
```

```
1040 FOR I=1 TO N

1050 READ W(I),DP(I)

1060 NEXT I

1070 RETURN

1080 INPUT"DO YOU WANT AN ESTIMATE OF CYCLONE EFFICIENCY(Y/N)";A$

1090 IF A$="N" THEN 590

1100 CLS

1110 PRINT"CHOOSE DUST CLASSIFICATION(1-3)"

1120 PRINT"1. COARSE
            2. FINE
            3. SUPER FINE"

1130 INPUT Z

1135 CLS

1140 FOR I1=1 TO 11

1150 READ W(I1),DP(I1)

1160 NEXT I1

1170 IF Z=1 THEN N=11:RESTORE:RETURN

1180 FOR I1=1 TO 12

1190 READ W(I1),DP(I1)

1200 NEXT I1

1210 IF Z=2 THEN N=12:RESTORE:RETURN

1220 FOR I1=1 TO 7

1230 READ W(I1),DP(I1)

1240 NEXT I1

1250 N=7:RESTORE:RETURN

3000 DATA .54,109,.06,67,.08,50,.05,35,.06,25,.05,17.5,.04,12.5,.03,9,.
      03,6.25,.03,3.75,.03,1.5

3010 DATA .03,125,.07,90,.10,67,.15,50,.10,35,.10,25,.07,17.5,.08,12.5,
      .04,9,.06,6.25,.08,3.75,.12,1.5

3020 DATA .05,67,.01,17.5,.04,12.5,.05,9,.10,6.25,.19,3.75,.56,1.5
```

LISTING FOR BASIC PROGRAM "CYCLONE2"

```
10 CLS:DEFINT I

15 Z$(1)="COARSE":Z$(2)="FINE":Z$(3)="SUPER FINE"

20 DIM DP(20),W(20)

30 PRINT"CYCLONE DESIGN PROGRAM"

40 INPUT"FLOW-SCFH:";Q

50 INPUT"PRESSURE-PSIA:";P

60 INPUT"TEMP-F:";T

70 INPUT"GAS VISCOSITY-CP:";CP

80 INPUT"PARTICLE DENSITY-LB/FT3:";PP

90 INPUT"GAS MOLECULAR WEIGHT:";MW

100 INPUT"ALLOWABLE PRESSURE DROP-IN. H2O:";HW

110 CLS

120 INPUT"DO YOU WANT AN ESTIMATE OF CYCLONE EFFICIENCY-(Y/N)";A$

130 IF A$="N" THEN 310

140 CLS:PRINT"CHOOSE DUST CLASSIFICATION"

150 PRINT"1. COARSE"

160 PRINT"2. FINE"

170 PRINT"3. SUPER FINE"

180 INPUT"CHOOSE CLASSIFICATION (1-3)";Z

190   GOSUB 1000

310 X1=(T+460)/460*14.7/P

320 PF=MW/(359*X1)

330 Q1=Q*X1/3600

340 DC=(1.92*PF*Q1[Z/HW)[.25

350 DE=DC/2

360 A=DC/2

370 B=DC*.2
```

```
380 S=DC/2

390 H=DC*1.5

400 HH=DC*4

410 BB=DC*.375

420 V1=Q1/(A*B)

430 V2=Q1/(DE[2/4*3.1412G)

440 W=((4*32.2*CP*(PP-PF)*.000674)/(3*PF[2))[.3333

450 X=(B/DC)/(B/DC)[.4

460 VS=2.055*W*X*DC[.067*V1[.667

470 X=(1-(12/DC)[.14/2.5)

480 Y=((T+460)/530)[.30

490 N=1-X*Y

500 CLS

510 PRINT"                    TABLE OF RESULTS"

520 PRINT"FLOW-SCFH=";Q;TAB(35);"PRESS-PSIA=";P

530 PRINT"TEMP-F=";T;TAB(35);"GAS VISCOSITY-CP=";CP

540 PRINT"MOLE WT=";MW;TAB(35);"SOLIDS DENSITY-LB/FT3=";PP

550 PRINT"ALLOWABLE PRESS DROP (IN/H2O)=";HW

580 PRINT TAB(20);"CALCULATED CYCLONE DIMENSIONS"

590 PRINT"INLET HEIGHT(FT)=";A;TAB(35);"INLET WIDTH(FT)=";B

600 PRINT"BARREL DIA(FT)=";DC;TAB(35);"OUTLET DIA(FT)=";DE

610 PRINT"BARREL LENGTH(FT)=";H;TAB(35);"OVERALL LENGTH(FT)=";HH

620 PRINT"SHAVE OFF(FT)=";S;TAB(35);"DUST OUTLET DIA(FT)=";BB

630 PRINT"INLET VEL(FT/SEC)=";V1;TAB(35);"OUTLET VEL(FT/SEC)=";V2

640 PRINT"INLET VELOCITY/SALTATION VELOCITY=";V1/VS

650 INPUT B$

660 CLS

670 IF A$="N" THEN 960
```

```
 680 IF Z=1 THEN I1=11

 690 IF Z=2 THEN I1=12

 700 IF Z=3 THEN I1=10

 710 CLS

 715 ET=0

 720 PRINT"ESTIMATED CYCLONE EFFICIENCY-";Z$(Z)

 730 PRINT"DP-MICRONS","WT-%","%-COLLECTED"

 740 FOR I=1 TO I1

 750 DP=DP(I)*3.25E-6

 760 TAU=PP*DP[2/(18*.000674*CP)

 770 X=551.3*TAU*Q1/DC[3*(N+1)

 780 Y=-2*X[(.5/(N+1))

 790 E=1-EXP(Y)

 800 ET=ET+W(I)*E

 810 PRINT DP(I),W(I),E*100

 815 NEXT I

 820 PRINT"OVERALL EFFICIENCY=";ET*100

 830 INPUT B$

 840 CLS

 850 INPUT"DO YOU WANT TO EXAMINE ANOTHER PARTICLE DISTRIBUTION(Y/N)";A
     $

 860 RESTORE

 930 IF A$="N" THEN 960

 950 GOTO 140

 960 INPUT"CHANGE CYCLONE PRESSURE DROP(Y/N)";P$

 965 IF P$="Y" THEN CLS:GOTO 100

 970 PRINT"PROGRAM EXECUTION TERMINATED"

 975 END

1000 FOR I=1 TO 11
```

calculating overall collection eff (handwritten annotation, pointing to lines 760–790)

```
1010 READ W(I),DP(I)

1020 NEXT I

1030 IF Z=1 THEN RETURN

1040 FOR I=1 TO 12

1050 READ W(I),DP(I)

1060 NEXT I

1070 IF Z=2 THEN RETURN

1080 FOR I=1 TO 10

1090 READ W(I),DP(I)

1100 NEXT I

1110 RETURN
2000 DATA .54,109,.06,67,.08,50,.05,35,.06,25,.05,17.5,.04,12.5,.03,9,.
     03,6.25,.03,3.75,.03,1.5
2010 DATA .03,125,.07,90,.10,67,.15,50,.10,35,.10,25,.07,17.5,.08,12.5,
     .04,9,.06,6.25,.08,3.75,.12,1.5
2020 DATA .05,67,0,50,0,35,0,25,.01,17.5,.04,12.5,.05,9,.10,6.25,.19,3.
     75,.56,1.50
```

VENTURI SCRUBBERS

Venturi wet scrubbers are widely used to remove low-micron and submicron particles from gas streams. Venturi scrubbers operating at high pressure drops (>50 in. water column) can achieve better than 90% collection efficiency of 0.1μ particles. Properly designed installations can be low in capital cost and economic to operate.

The program VENTURI can be used to design new scrubbers, to check the performance of existing systems, or to check the effect of changing operating conditions. The program requires the following information:

1. Venturi throat diameter
2. Gas flow rate
3. Liquid flow rate
4. Gas viscosity
5. Liquid viscosity
6. Liquid density
7. Particle density

8. Liquid surface tension
9. Particle size distribution

The program estimates collection efficiency by the equations proposed by Johnstone,[1] corrected for the effect of Brownian diffusion in the throat. Calvert[2] has proposed a more complicated model to estimate collection efficiency of venturi scrubbers. Calvert's model suggests that L/D ratios of venturi throats should be 2–3. Longer throats do not improve collection efficiency significantly and reduce pressure recovery in the diffuser section. As a practical matter, wet venturi scrubbers should be operated with liquid rates of 7–10 gpm/1000 cfm of gas flow. Pressure drops should be at least 10–15 in. water column to remove 1–2μ particles effectively. Throat velocities can range from 150 to 400 ft/sec. Higher throat velocities can cause problems such as erosion in the venturi throat. [Figure 7.3 is an illustration of a typical wet scrubber system. The included angle at the throat of the venturi should be 20°. The included angle in the diffuser section should be 10° to keep pressure recovery high.]

If the venturi is to operate with a gas other than air, the mean free path (MFP) must be changed in line 10. The mean free path can be estimated using the kinetic theory of viscosity of gases or looked up in a handbook. Johnstone reported that the value of K in line 10 can very from 750–1500. If no other information is available, use a value of 1000–1300 for K. The errors introduced are approximately ±20% for particles having low collection efficiency if the error in estimating K is approximately 50%.

Equations for VENTURI [3]

$$D_o = (18506/G_1) (\sigma/\rho_L)^{1/2}/V_g + 2825 (1000 * R)^{1.5} [\mu_L/(\sigma * \rho_L)^{1/2}]^{.45}$$

$$A1 = 1.257 + .4 * \exp(-1.1 Dp/2\lambda)$$

$$C = 1 + (2A1\lambda/Dp)$$

$$\psi = (C\rho_p V_g D_p^2)/(18 D_o \mu_g)$$

$$\eta_i = 1 - \exp(-K * R * \psi^{.5})$$

$$\eta_T = \sum_{i=1}^{n} w_i \eta_i$$

$$\Delta P = 10.22 V_g^2 R/2.52$$

$$A = (D^2/4 * \pi)/10.758$$

$$L_1 = L * 60 * .0038$$

$$G_1 = G/35.28$$

$$R = L_1/G_1$$

$$V_g = G_1/(A * 3600)$$

[1] Johnstone, H., *Ind. Eng. Chem. 46*, 1601 (1954).
[2] Calvert S.; Chow, S., Barbarird, H., *Envir. Sci. Tech. 12*, 456–458. (1978).
[3] Ollero, P. *Chem. Eng.*, May 28, 1984, p. 103.

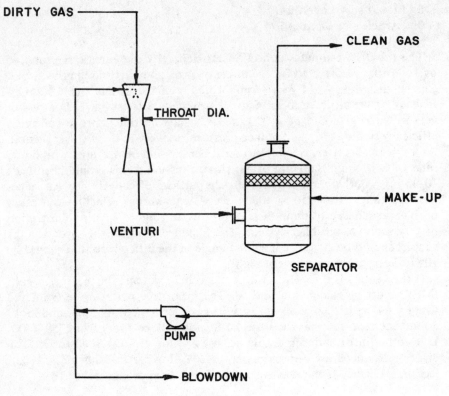

Figure 7.3

Nomenclature for VENTURI

VARIABLE	*DESCRIPTION*
G	Gas flow (ACFH) (actual cubic ft per hour)
L	Liquid flow (gpm) (gallons per minute)
D	Venturi diameter (ft)
ρ_L	Liquid density (Kg/m³)
ρ_p	Particle density (Kg/m³)
μ_L	Liquid viscosity (P) (cp * .01)
μ_g	Gas viscosity (Kg/m −s) (cp * .001)
D_p	Particle diameter (m)
D_o	Liquid drop diameter (m)
σ	Surface tension (dyn/cm)
ΔP	Pressure drop (in H₂0)
A	Flow area (m²)

VARIABLE	DESCRIPTION
L_1	Liquid rate (m³/hr)
G_1	Gas rate (m³/hr)
V_g	Gas velocity (m/sec)
R	Liquid gas ratio (m³ liq/m³ gas)
η_i	Fractional efficiency
η_T	Total efficiency
w_i	Weight % (decimal)
C	Correction factor
Al	Calculated constant
λ	Mean free path (m)
K	Venturi constant 750–1500

EXAMPLE

READY

RUN

VENTURI PERFORMANCE MODEL

OLD VENTURI DIA. (FT)=0

VENTURI THROAT DIA. (FT)=? **1**

OLD GAS RATE (CFH)=0

GAS RATE (CFH)=? **450000**

OLD LIQUID RATE (GPM)=0

LIQUID RATE (GPM)=? **100**

OLD GAS VISCOSITY (CP)=0

GAS VISCOSITY (CP)=? **.0223**

LIQUID VISCOSITY (CP)=0

LIQUID VISCOSITY (CP)=? **1**

OLD LIQUID DENSITY (GRAM/ML)=0

LIQUID DENSITY (GR/CC)=? **1**

OLD PRACTICE DENSITY (GRAM/ML)=0

PARTICLE DENSITY (GRAM/ML)=? **2**

OLD LIQUID SURFACE TENSION (DYNES/CM)=0
LIQUID SURFACE TENSION (DYNES/CM)= **72**

HOW MANY PARTICLE SIZES:? **6**
N=1
DP-MICRONS=? **.10**
WT% (1=1%)=? **1**

N=2
DP-MICRONS=? **.2**
WT% (1=1%)=? **10**

N=3
DP-MICRONS=? **.50**
WT% (1=1%)=? **20**

N=4
DP-MICRONS=? **.8**
WT% (1=1%)=? **20**

N=5
DP-MICRONS=? **1**
WT% (1=1%)=? **25**

N=6
DP-MICRONS=? **2**
WT% (1=1%)=? **24**

DP-MICRONS	WT%	EFF%	%ESCAPING
.1	1	30.2178	69.7822
.2	10	44.0701	55.9299
.5	20	70.6327	29.3673

DP-MICRONS	WT%	EFF%	%ESCAPING
.8	20	84.5315	15.4685
1	25	89.9019	10.0981
2	24	98.7978	1.2022

ESTIMATED OVERALL EFFICIENCY = 81.929%

?

VENTURI DIA (FT) = 1   THROAT VELOCITY (FT/SEC) = 159.093

LIQUID RATE (GPM) = 100   GAS RATE (CFH) = 450000

LIQ. VISC. (CP) = 1   GAS VISC. (CP) = .0223

LIQUID DENSITY (GR/CC) = 1   PARTICLE DENSITY (GR/CC) = 2

SURFACE TENSION (DYNES/CM) = 72

ESTIMATED PRESSURE DROP (IN H20) = 17.0551

?

MAKE ANY CHANGES (Y/N)?   N

READY


```
    LISTING FOR BASIC PROGRAM "VENTURI"

 10 CLS:PI=3.1416:K=1000:MFP=6.53E-8

 20 PRINT

 30 GOSUB 490

 40 GOSUB 530

 50 GOSUB 570

 60 GOSUB 620

 70 GOSUB 670

 80 GOSUB 720

 90 GOSUB 770

100 GOSUB 820

110 GOSUB 1060
```

```
120 GOSUB 970

130 ET=0

140 CLS

150 PRINT"Dp-MICRONS","WT%","EFF%","% ESCAPING"

160 FOR I=1 TO N

170 DP=DP(I)/1E06

180 A1=1.257+.4*EXP(-1.1*DP/(2*MFP))

190 C=1+(2*MFP*A1/DP)

200 SI=C*PS*VG*DP[2/(18*D0*UG)

210 E=1-EXP(-K*R*SI[.5)

250 ET=ET+E*W(I)

260 PRINT DP(I),W(I)*100,E*100,(1-E)*100

270 NEXT I

280 PRINT"ESTIMATED OVERALL EFFICIENCY=";ET*100

290 INPUT A$

300 CLS

310 PRINT"VENTURI DIA(FT)=";D;TAB(30);"THROAT VELOCITY(FT/SEC)=";VG*3.
    28

320 PRINT"LIQUID RATE(GPM)=";L;TAB(30);"GAS RATE(CFH)=";G

330 PRINT"LIQ. VISC.(CP)=";UL*100;TAB(30);"GAS VISC.(CP)=";UG*1000

340 PRINT"LIQUID DENSITY(GRAM/ML)=";P1L;TAB(30);"PARTICLE DENSITY(GRAM
    /ML)=";XP

350 PRINT"SURFACE TENSION(DYNES/CM)=";SG

360 PDROP=10.22*VG[2*R

370 PRINT"ESTIMATED PRESSURE DROP(IN H2O)=";PDROP/2.52

380 INPUT A1$

390 CLS

400 INPUT"MAKE ANY CHANGES(Y/N)";A$

410 IF A$="N" THEN 480
```

```
420 GOSUB 870

430 PRINT

440 INPUT"CHANGE WHICH VARIABLE(0-8, ENTER 0 TO END)";Z

450 IF Z=0 THEN 120

460 ON Z GOSUB 490  ,530  ,570  ,620  ,670  ,720  ,770  ,820

470 GOTO 420

480 END

490 CLS:PRINT"VENTURI PERFORMANCE MODEL":PRINT"OLD VENTURI DIA.(FT)=";
    D

500 PRINT

510 INPUT"VENTURI THROAT DIA.(FT)=";D

520 RETURN

530 CLS:PRINT"OLD GAS RATE(CFH)=";G

540 PRINT

550 INPUT"GAS RATE(CFH)=";G

560 RETURN

570 CLS:PRINT"OLD LIQUID RATE(GPM)=";L

580 PRINT

590 INPUT"LIQUID RATE(GPM)=";L

600 RETURN

610 CLS

620 CLS

630 PRINT"OLD GAS VISCOSITY(CP)=";V1G

640 PRINT

650 INPUT"GAS VISCOSITY(CP)=";V1G

660 RETURN

670 CLS

680 PRINT"LIQUID VISCOSITY(CP)=";V2L

690 PRINT
```

```
700 INPUT"LIQUID VISCOSITY(CP)=";VZL

710 RETURN

720 CLS

730 PRINT"OLD LIQUID DENSITY(GRAM/ML)=";P1L

740 PRINT

750 INPUT"LIQUID DENSITY(GRAM/ML)=";P1L

760 RETURN

770 CLS

780 PRINT"OLD PARTICLE DENSITY(GRAM/ML)=";XP

790 PRINT

800 INPUT"PARTICLE DENSITY(GRAM/ML)=";XP

810 RETURN

820 CLS

830 PRINT"OLD LIQUID SURFACE TENSION(DYNES/CM)=";SG

840 PRINT

850 INPUT"LIQUID SURFACE TENSION(DYNES/CM)=";SG

860 RETURN

870 CLS

880 PRINT"1. VENTURI THROAT DIAMETER"

890 PRINT"2. GAS FLOW RATE"

900 PRINT"3. LIQUID FLOW RATE"

910 PRINT"4. GAS VISCOSITY"

920 PRINT"5. LIQUID VISCOSITY"

930 PRINT"6. LIQUID DENSITY"

940 PRINT"7. PARTICLE DENSITY"

950 PRINT"8. LIQUID SURFACE TENSION"

960 RETURN

970 G1=G/35.28
```

```
980 L1=L*60*.0038

990 R=L1/G1

1000 A=D[2/4*PI/10.758

1010 VG=G/(35.3*3600*A)

1020 UG=V1G*.001:UL=VZL*.01:PL=P1L*1000:PS=XP*1000

1030 D0=18506*(SG/PL)[.5/VG+2825*(1000*R)[1.5*(UL/(SG*PL)[.5)[.45

1040 D0=D0/1E6

1050 RETURN

1060 CLS

1070 INPUT"HOW MANY PARTICLE SIZES:";N

1080 FOR I=1 TO N

1090 PRINT "N=";I

1100 INPUT"Dp-MICRONS=";DP(I)

1110 INPUT"WT%(1=1%)=";W(I)

1120 W(I)=W(I)/100

1130 CLS

1140 NEXT I

1150 RETURN
```

PACKED TOWER DESIGN

Packed towers have become popular equipment choices for distillation, absorption, and stripping operations. New packing materials have been developed that offer low pressure drop per theoretical tray and low theoretical plate heights.

The calculations involved in determining the height of packing are complex. The calculations performed by the program ABSORBER include material balances, interfacial concentrations, number of overall gas transfer units, height of the overall gas transfer unit, and tower height. The program breaks the Y1 − Y2 difference down to between 20–50 intervals. The program then integrates the area under the curve, dY/(Y − Y*), for each interval. Interfacial concentrations are calculated for each interval and then are used to calculate the height of the overall gas transfer unit. The number of gas transfer units is summed over all the intervals and multiplied by the current height of a transfer unit to determine the height of the tower.

The functions for the equilibrium curve should be added starting at line 850. The functions for the heights of the liquid and gas transfer units are added at line 980. You do not need a function for the equilibrium curve or the transfer units to use the program. If these variables will remain constant, simply type RETURN at the appropriate line in subroutines starting after lines 850 and 980.

The program reads initial data from data statements starting at line 1300. For absorbers, the data must be arranged as follows:

1300 DATA NAME,FLAG

1310 DATA Y1,Y2,X2,G1,L2

1320 DATA S,HG,HL,RPHI,Z1

For absorbers, FLAG=0.
For strippers, the data must be arranged as follows:

1300 DATA NAME,FLAG

1310 DATA X1,X2,Y2,G2,L1

1320 DATA S,HG,HL,RPHI,Z1

For strippers, FLAG=1.

Although the program is designed to handle isothermal columns, columns with small temperature rises ($10-20°F$) can be accommodated by building in a heat balance at subroutine 850 to calculate the liquid temperature rise. The key assumption here is that the sensible heat effects from warming or cooling the gas are small compared to the heat of solution. For columns in which heat effects are large, you can program the differential equations for adiabatic towers as stated in Sherwood and Wilke *Mass Transfer* (1975), pp., 489–497, and solve them using DIFFEQ. This will account for all heat effects and allow the construction of the equilibrium curve as a function of liquid composition throughout the tower. The equilibrium curve is then programmed into subroutine 850. ABSORBER will correctly calculate the depth of packing required.

Sherwood, T., Wilke, R., "Mass Transfer" McGraw-Hill, NY, NY, 1975, pg. 489–497, Appendix A.

Figure 7.4 provides the nomenclature needed to provide ABSORBER with the input data the program needs.

Equations for ABSORBER

$$NT = \int DY/(Y - Y^*) + .5 * LN [(1 - Y2)/(1 - Y1)]$$
$$HOG = HG (1 - Y_i)_{lm}/(1 - Y)_{lm} + S * G/L * HL (1 - X_i)_{lm}/(1 - X)_{lm}$$
$$TL = NT * HOG$$

For absorbers
$$GC = (1 - Y1 - Z1) * G1$$
$$G0 = (1 - Y1 * RPHI) * G1$$
$$L0 = (1 - X2 * RPHI) * L2$$

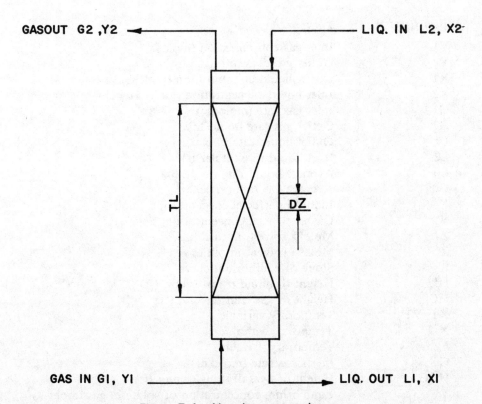

Figure 7.4 Absorber nomenclature.

For strippers

$$GC = (1 - Y2 - Z2) * G2$$
$$G0 = (1 - Y2 * RPHI) * G2$$
$$L0 = (1 - X1 * RPHI) * L1$$
$$L2 = L0/(1 - X2 * RPHI)$$
$$XT = 1 - X1 * RPHI$$
$$Y1 = L0 * X1 + G2 * Y2 * XT - L2 * X2 * XT$$
$$Y1 = Y1/[G2 * XT + L2 * RPHI * (X1 - X2)]$$

Overall material balances (absorbers and strippers)

$$YT = 1 - RPHI * Y$$
$$X = G0 * Y - G2 * Y * YT + L2 * X2 * YT$$
$$X = X/[L2 * YT + G2 * RPHI * (Y - Y2)]$$

Nomenclature for ABSORBER

VARIABLE	DESCRIPTION
Y1	Inlet gas concentration (mole%)
Y2	Outlet gas concentration (mole%)
X1	Outlet liquid concentration (mole%)
X2	Inlet liquid concentration (mole%)
G1	Inlet gas rate (mole/hr)
G2	Outlet gas rate (mole/hr)
L1	Outlet liquid rate (mole/hr)
L2	Inlet liquid rate (mole/hr)
L0	Average liquid rate (mole/hr)
G0	Average gas rate (mole/hr)
X_i	Liquid interfacial concentration (mole%)
Y_i	Gas interfacial concentration (mole%)
Z1	Mole% solvent in inlet gas
Z2	Mole% solvent in outlet gas
S	Slope of equilibrium curve
HL	Height of liquid transfer unit (ft)
HG	Height of gas transfer unit (ft)
XT, YT	Temporary variable
X	Temporary variable
Y	Temporary variable
GC	Inerts gas rate (mole/hr)
HOG	Height of overall transfer unit (ft)
Y*	Equilibrium concentration of solute in gas (mole%)
TL	Tower height (ft)

VARIABLE	DESCRIPTION
lm	Log mean difference
NT	Number of overall gas transfer units
RPHI	1-average volume % solvent in gas throughout the tower (decimal)

EXAMPLE

PACKED TOWER DESIGN PROGRAM

HOW MANY INTEGRATION INTERVALS (20-50)? **20**

NO. OF INTERVALS=20

SYSTEM TITLE-TEST

$Y1 = .02$ $Y2 = 1.2E - 03$ $X1 = .106216$ $X2 = 5E - 03$

$G1 = 80.8$ $L2 = 13.43$

$S = .125$ $HG = 1.735$ $HL = 1.61$

HG	HL	HOG	NT	TL
1.735	1.610	2.8047	0.1329	0.3726
1.735	1.610	2.8024	0.2731	0.7655
1.735	1.610	2.8079	0.4213	1.1818
1.735	1.610	2.8135	0.5785	1.6241
1.735	1.610	2.8191	0.7456	2.0951
1.735	1.610	2.8245	0.9238	2.5984
1.735	1.610	2.8304	1.1144	3.1379
1.735	1.610	2.8359	1.3191	3.7183
1.735	1.610	2.8416	1.5397	4.3454
1.735	1.610	2.8475	1.7788	5.0262
1.735	1.610	2.8533	2.0393	5.7693
1.735	1.610	2.8592	2.3247	6.5855
1.735	1.610	2.8650	2.6400	7.4887
1.735	1.610	2.8710	2.9911	8.4968
1.735	1.610	2.8766	3.3865	9.6341

HG	HL	HOG	NT	TL
1.735	1.610	2.8827	3.8374	10.9341
1.735	1.610	2.8887	4.3605	12.4451
1.735	1.610	2.8967	4.9806	14.2414
1.735	1.610	2.9009	5.7383	16.4394
1.735	1.610	2.9069	6.7060	19.2524
1.735	1.610	2.9132	8.0341	23.1151
			8.0436	23.1428

LISTING FOR BASIC PROGRAM "ABSORBER"

```
10 F$="##.###      ##.###      ##.####     ##.####     ##.####    "

20 F1$="                                   ##.####    ###.####"

30 DIM Y(60)

40 CLS

50 PRINT"PACKED TOWER DESIGN PROGRAM"

60 INPUT"HOW MANY INTEGRATION INTERVALS(20-50)";IN

70 IF IN<20 OR IN>50 THEN PRINT"TRY AGAIN":GOTO 60

80 READ KT$,FLAG

90 IF FLAG<>0 THEN GOSUB 870

100 IF FLAG<>0 THEN 170

110 READ Y1,Y2,X2,G1,L2

120 READ S,HG,HL,RPHI,Z1

130 GC=(1-Y1-Z1)*G1

140 G0=(1-Y1*RPHI)*G1

150 L0=(1-X2*RPHI)*L2

160 GZ=G0/(1-YZ*RPHI)

170 REM-SET UP GRID FOR INTEGRATION

180 DY=(Y1-Y2)/IN
```

```
190 JE=(Y1-YZ)/DY

200 YZ=Y1

210 FOR J=1 TO JE+1

220 Y(J)=YZ

230 YZ=YZ-DY

240 NEXT J

250 JT=JE+1

260 Y(JT)=YZ

270 LPRINT"SYSTEM TITLE=";KT$

275 LPRINT

280 IF FLAG<>0 THEN 340

290 Y=Y1:GOSUB 800   :X1=X

300 LPRINT"Y1=";Y1,"YZ=";YZ,"X1=";X1,"XZ=";XZ

305 LPRINT

310 LPRINT"G1=";G1,"LZ=";LZ

315 LPRINT

320 LPRINT"S=";S,"HG=";HG,"HL=";HL

330 GOTO 360

340 LPRINT"X1=";X1,"XZ=";XZ,"YZ=";YZ:"Y1=";Y1

345 LPRINT

350 LPRINT"S=";S,"HG=";HG,"HL=";HL

360 REM-BEGIN TOWER HT COMPUTATIONS

370 LPRINT

380 LPRINT"HG          HL          HOG          NT          TL"

385 LPRINT

390 TL=0:NT=0

400 FOR J=1 TO JE+1

410 F1=0
```

```
420 DELY=DY/40:YH=Y(J)

430 REM-CALCULATE INTERFACIAL CONCENTRATIONS

440 GOSUB 800

450 XH=X

460 IF HL=0 THEN HOG=HG:GOTO 585

470 XI=.90*XH:TEMP=XI:DX=XH/10

480 XI=XI+DX

490 GOSUB 1000

500 F0=F

510 XI=XI-DX

520 GOSUB 1000

530 F2=F

540 DRV=(F0-F2)/DX

550 XI=XI-F2/DRV

560 IF ABS((TEMP-XI)/XI)<=.01 THEN 585

570 TEMP=XI

580 GOTO 480

584 REM-INTEGRATE AREA UNDER CURVE

585 FOR N=1 TO 40

590 Y=YH:GOSUB 800  :GOSUB 850  :F3=1/(YH-S*X)

600 IF N>1 AND N<=39 THEN F3=2*F3

610 YH=YH+DELY:F1=F1+F3

620 NEXT N:IF HL=0 THEN 690

630 GOSUB 1070 'CALCULATE LOG MEAN DIFFERENCES

640 X=XH

650 GOSUB 850

660 GOSUB 980

670 HOG=HG*LYI/LMY+S*G0/L0*HL*LXI/LMY

680 PRINT YI,XI
```

```
690 N1=DELY*F1/2:NT=NT+N1

700 DH=N1*HOG

710 TL=TL+DH

720 LPRINT USING F$;HG;HL;HOG;NT;TL

725 LPRINT

730 NEXT J

740 NT=NT+.5*LOG((1-Y2)/(1-Y1))

750 TL=TL+HOG*.5*LOG((1-Y2)/(1-Y1))

760 LPRINT

770 LPRINT USING F1$;NT;TL

780 PRINT"PROGRAM EXECUTION TERMINATED"

790 END

800 REM-CALCULATE MATERIAL BALANCE

810 YT=1-RPHI*Y

820 X=G0*Y-G2*Y2*YT+L2*X2*YT

830 X=X/(L2*YT+G2*RPHI*(Y-Y2))

840 RETURN

850 REM-INSERT EQUILIBRIUM FUNCTION HERE

860 RETURN

870 REM-READ IN DATA FOR A STRIPPER

880 READ X1,X2,Y2,L1,G2

890 READ S,HG,HL,RPHI,Z1

900 GC=(1-Y2-Z1)*G2

910 G0=(1-Y2*RPHI)*G2

920 L0=(1-X1*RPHI)*L1

930 L2=L0/(1-X2*RPHI)

940 XT=1-RPHI*X1

950 Y1=L0*X1+G2*Y2*XT-L2*X2*XT

960 Y1=Y1/(G2*XT+L2*RPHI*(X1-X2))
```

```
 970 RETURN

 980 REM-CALCULATE HT OF A TRANSFER UNIT

 990 RETURN

1000 REM-CALCULATE INTERFACIAL CONC FUNCTION

1010 X=XI

1020 GOSUB 850

1030 X3=(L0*HG)/(G0*HL)

1040 X4=((1-XH)/(1-XI))[X3

1050 F=(1-YH)*X4+XI*S-1

1060 RETURN

1070 REM-CALCULATE LOG MEAN DIFFERENCES

1080 X=XH

1090 GOSUB 850

1100 D1=(1-XH*S)-(1-YH)

1110 R1=LOG((1-XH*S)/(1-YH))

1120 LMY=D1/R1

1130 X=XI

1140 GOSUB 850

1150 YI=XI*S

1160 D1=(1-YI)-(1-YH)

1170 R1=LOG((1-YI)/(1-YH))

1180 LYI=D1/R1

1190 D1=(1-XI)-(1-XH)

1200 R1=LOG((1-XI)/(1-XH))

1210 LXI=D1/R1

1220 RETURN

1300 DATA TEST,0

1310 DATA .02,.0012,.005,80.8,13.43

1320 DATA .1250,1.735,1.61,1,0
```

8
MISCELLANEOUS

LIQUID AND GAS ORIFICE METERS

ORIFICE is a program that will calculate the bore of an orifice, venturi, or flow nozzle. The program can handle liquids or gases. If the gas compressibility factor is known, the program will take it into account. Gas expansion ratios are calculated automatically.

You can also use the program to determine the range of a given device if you know the bore, pipe ID, and meter differential. The program is self-documenting, and all user input is entered through the keyboard. All of the primary device equations are based on equations from Spink, *Principles and Practice of Flowmeter Engineering* (1973).[1]

Equations for ORIFICE

$S = W/[2835 * D^2 * (GF * HW)^{1/2} * FC]$ For liquids

$GF = GL/62.3$

$S = W/[359 * D^2 * Y * (PG * HW)^{1/2}]$ For gases

$PG = MW/(PF * 10.73 * TF)$

$X = d/D$

$Y = 1 - [.333 + 1.145 * (X^2 + .7 * X^5 + 12 * X^{13})] * X1/K$
    For flange taps

$Y = 1 - (.41 + .35 * X^4) * X1/K$ For all other tap arrangements

$X1 = HW/(27.7 * PF)$

$F1 = .598 * X^2 + .01 * X^3 + 1.947E - 5 * X^2 * (10 * X)^{4.425} - S$
    (Flange taps)

---

[1]Spink, L., *The Principles and Practice of Flow Meter Engineering,* 9th ed., The Foxboro Company, Foxboro, Mass., 1973

$$F1 = .58925 * X^2 + .272 * X^3 - .825 * X^4 + 1.75X^5 - S$$
$2\frac{1}{2} - 8D$ taps

$$F1 = .98 * [X^4/(1 - X^4)]^{1/2} - S \qquad \text{Flow Nozzles and Venturis}$$

$$A = (D/12)^2 * \pi/4$$

$$G1 = W/A$$

$$RD = (D/12) * G1/(2.42 * CP)$$

$$FC = 1$$

Nomenclature for ORIFICE

VARIABLE	*DESCRIPTION*
X	Beta ratio d/D
S	Meter constant
W	Flow rate (lb/hr)
D	Pipe ID (in.)
GL	Liquid density (lb/ft^3)
GF	Specific gravity
PG	Gas density (lb/ft^3)
Y	Expansion factor
PF	Flowing gas pressure (psia)
TF	Flowing temperature (°R) Degrees Rankine
MW	Molecular weight of gas
K	C_p/C_v ratio of gas
CP	Fluid viscosity (CP)
X1	Pressure drop correction for gases
F1	Device bore function
A	Pipe flow area (ft^2)
G1	Superficial mass flow rate (lb/hr-ft^2)
RD	Pipe reynolds number
FC	Viscosity correction factor
π	3.1416
C_p	Gas heat capacity of constant pressure (BTU/LB-°F)
C_v	Gas heat capacity at constant volume(BTU/LB-°F)
d	Device bore (inches)

EXAMPLE

READY

RUN

FLOW METER PROGRAM

LIQUID (L) OR GAS (G) ?   **L**

FLUID NAME=?   **WATER**

DEVICE TAG=?   **FE-100**

CALCULATE BORE(1) OR RANGE OR A DEVICE(2)?   **1**

MAX FLOW-PPH=?   **10000**

METER DIFFERENTIAL (IN. H20)=?   **100**

METER RUN ID(INCHES)=?   **3.068**

FLUID VISCOSITY(CP)=?   **1**

LIQUID DENSITY (LB/FT3)=?   **62.3**

WHAT TYPE OF METER?

1. ORIFICE METER-FLG, VENA CONTRACTA, RADIUS, OR CORNER
TAPS

2. ORIFICE METER-2½ − 8D TAPS

3. FLOW NOZZLE OF VENTURI TUBE

METER CHOICE (1 − 3)=?   **1**

DEVICE TAG=FE-100

FLUID NAME=WATER

PRIMARY DEVICE=ORIFICE-FLG, VC, RAD, CT TAPS

MAX FLOW-PPH=10000   HW(INCHES)=100

METER RUN ID(INCHES)=3.068   BORE(INCHES)=.765713

FLUID VISCOSITY(CP)=1   SPECIFIC GRAVITY=1

S=.0374745   RE=20578.8   FC=1

?

CALCULATE BORE FOR ANOTHER DEVICE(Y/N)?   **N**

PROGRAM EXECUTION TERMINATED

READY

EXAMPLE

RUN

FLOW METER PROGRAM

LIQUID (L) OR GAS (G) ?   **G**

FLUID NAME=?   **AIR**

DEVICE TAG=?   **FE-200**

CALCULATE BORE(1) OR RANGE OR A DEVICE(2)?   **1**

MAX FLOW-PPH=?   **5000**

METER DIFFERENTIAL (IN. H20)=?   **100**

METER RUN ID(INCHES)=?   **3.068**

FLUID VISCOSITY(CP)=?   **.0125**

GAS MOLE WT.=?   **29**

FLOWING TEMP (R)=?   **520**

FLOWING PRESSURE (PSIA)=?   **24.7**

COMPRESSIBILITY FACTOR-Z=?   **1**

CP/CV RATIO(AIR=1.39)=?   **1.391**

WHAT TYPE OF METER?

1. ORIFICE METER-FLG, VENA CONTRACTA, RADIUS, OR CORNER TAPS

2. ORIFICE METER-2-1/2 − 8D TAPS

3. FLOW NOZZLE OF VENTURI TUBE

METER CHOICE (1 − 3)=?   **1**

DEVICE TAG=FE-200

FLUID NAME=AIR

PRIMARY DEVICE=ORIFICE-FLG, VC, RAD, CT TAPS

MAX FLOW-PPH=5000	HW(INCHES)=100
METER RUN ID(INCHES)=3.068	BORE(INCHES)=2.32816
FLUID VISCOSITY(CP)=.0125	SPECIFIC GRAVITY=1
GAS MW=29	FLOWING TEMP(R)=520
PRESSURE (PSIA)=24.7	Z-FACTOR=1

CP/CV RATIO=1.391

$S=.436711$ $Re=823151$ $FC=1$ $Y=.945638$

?

CALCULATE BORE FOR ANOTHER DEVICE(Y/N)?   N

PROGRAM EXECUTION TERMINATED

READY

EXAMPLE

RUN

FLOW METER PROGRAM

LIQUID (L) OR GAS (G) ?   L

FLUID NAME = ?   WATER

DEVICE TAG = ?   FE-300

CALCULATE BORE(1) OR RANGE OR A DEVICE(2)?   2

METER DIFFERENTIAL(IN. H20) = ?   200

METER RUN ID(INCHES) = ?   3.068

FLUID VISCOSITY(CP) = ?   1

LIQUID DENSITY (LB/FT3) = ?   62.3

WHAT TYPE OF METER?

1. ORIFICE METER-FLG, VENA CONTRACTA, RADIUS, OR CORNER
TAPS

2. ORIFICE METER-2½ − 8D TAPS

3. FLOW NOZZLE OF VENTURI TUBE

METER CHOICE (1 − 3) = ?   2

BORE OF DEVICE(INCHES) = ?   1.500

DEVICE TAG = FE-300

FLUID NAME = WATER

PRIMARY DEVICE = ORIFICE-2-1/2-8D TAPS

MAX FLOW-PPH = 65812.3	HW(INCHES) = 200
METER RUN ID(INCHES) = 3.068	BORE(INCHES) = 1.5
FLUID VISCOSITY(CP) = 1	SPECIFIC GRAVITY = 1

S = .174393   Re = 135434   FC = 1

?

PROGRAM EXECUTION TERMINATED

READY

   LISTING FOR BASIC PROGRAM "ORIFICE"

```
10 A1$(1)="ORIFICE-FLG, VC, RAD, CT":A1$(2)="ORIFICE- 2-1/2 -8D":A1$(
   3)="FLOW NOZZLE OR VENTURI TUBE"

20 CLS

30 PRINT"FLOW METER PROGRAM":N=2835:PI=3.1416

40 INPUT"LIQUID(L) OR GAS(G)";A$

50 INPUT"FLUID NAME=";F$

60 INPUT"DEVICE TAG=";TAG$:CLS

70 INPUT"CALCULATE BORE(1) OR RANGE OF A DEVICE(2)";A1

80 IF A1=2 THEN 100

90 INPUT"MAX FLOW-PPH=";W

100 INPUT"METER DIFFERENTIAL(IN. H2O)=";HW

110 INPUT"METER RUN ID(INCHES)=";D

120 INPUT"FLUID VISCOSITY(CP)=";CP

130 IF A$="L" THEN 200

140 INPUT"GAS MOLE WT.=";MW

150 INPUT"FLOWING TEMP(R)=";TF

160 INPUT"FLOWING PRESSURE(PSIA)=";PF

170 INPUT"COMPRESSIBILITY FACTOR-Z=";Z

180 INPUT"CP/CV RATIO(AIR=1.39)=";K

190 GOTO 210

200 INPUT"LIQUID DENSITY(LB/FT3)=";GL

210 CLS
```

```
220 PRINT"WHAT TYPE OF METER?"

230 PRINT"1. ORIFICE METER-FLG, VENA CONTRACTA, RADIUS, OR CORNER TAPS
    "

240 PRINT"2. ORIFICE METER 2-1/2 -8D TAPS"

250 PRINT"3. FLOW NOZZLE OR VENTURI TUBE"

260 INPUT"METER CHOICE (1-3)=";M

270 CLS

280 IF A1=2 THEN GOSUB 940   :GOTO 450

290 CLS

300 IF A$="G" THEN 360

310 GF=GL/62.3:TEMP=1.:FC=1.

320 GOSUB 650

330 S=W/(N*D[2*SQR(GF*HW)*FC)

340 GOSUB 680 · 'FIND BORE

350 GOTO 450

360 REM-BEGIN GAS CALCULATIONS

370 GOSUB 650   :Y=1.0:FC=1.0:GF=MW/29:TEMP=1.0

380 PG=MW*PF/(10.73*TF*Z)

390 FOR J=1 TO 5:S=W/(359*D[2*FC*Y*SQR(PG*HW))

400 GOSUB 680

410 X1=HW/(27.7*PF)

420 GOSUB 850

430 REM-ITERATE UNTIL Y VALUE EXCEEDS 4 PLACE ACCURACY

440 NEXT J

450 CLS:REM- PRINT RESULTS

460 PRINT"---------------------------------------------------------"

470 PRINT"DEVICE TAG=";TAG$

480 PRINT"FLUID NAME=";F$

490 PRINT"---------------------------------------------------------"
```

```
500 PRINT"PRIMARY DEVICE=";A1$(M);" TAPS"

510 PRINT"---------------------------------------------------------"

520 PRINT"MAX FLOW-PPH=";W;TAB(30);"HW(INCHES)=";HW

530 PRINT"METER RUN ID(INCHES)=";D;TAB(30);"BORE(INCHES)=";X*D

540 PRINT"FLUID VISCOSITY(CP)=";CP;TAB(30);"SPECIFIC GRAVITY=";GF

550 IF A$="G" THEN GOSUB 900

560 PRINT"S=";S;"     ";"Re=";RD;"     ";"FC=";FC;"     ";

570 IF A$="G" THEN PRINT"Y=";Y ELSE PRINT

580 INPUT Z$:CLS

590 IF A1=2 THEN 620

600 INPUT"CALCULATE BORE FOR ANOTHER DEVICE(Y/N)";Z$

610 IF Z$="Y" THEN 210

620 CLS

630 PRINT"PROGRAM EXECUTION TERMINATED"

640 END

650 A=PI*D[2/576:G1=W/A

660 RD=D/12*G1/(2.42*CP)

670 RETURN

680 XTEMP=.6:X=.6

690 X=X+.0001

700 ON M GOSUB 790 ,810 ,830

710 F0=F1

720 X=X-.0001

730 ON M GOSUB 790 ,810 ,830

740 DERV=(F0-F1)/.0001

750 X=X-F1/DERV

760 IF ABS(XTEMP-X)=<.0001 THEN RETURN

770 XTEMP=X
```

```
780 GOTO 690

790 F1=.598*X[2+.01*X[3+1.947E-5*X[2*(10*X)[4.425-S

800 RETURN

810 F1=.58925*X[2+.272*X[3-.825*X[4+1.75*X[5-S

820 RETURN

830 F1=.98*SQR(X[4/(1-X[4))-S

840 RETURN

850 IF M=1 THEN 880

860 Y=1-(.333+1.145*(X[2+.7*X[5+12*X[13))*X1/K

870 GOTO 890

880 Y=1-(.41+.35*X[4)*X1/K

890 RETURN

900 PRINT"GAS MW=";MW;TAB(30);"FLOWING TEMP(R)=";TF

910 PRINT"PRESSURE(PSIA)=";PF;TAB(30);"Z-FACTOR=";Z

920 PRINT"CP/CV RATIO=";K

930 RETURN

940 REM-SUBROUTINE TO CALCULATE THE RANGE OF A DEVICE

950 FC=1.0

960 GF=GL/62.3:IF A$="G" THEN PG=MW*PF/(10.73*TF*Z):GF=MW/29

970 INPUT"BORE OF DEVICE(INCHES)=";D1

980 X=D1/D

990 S=0:ON M GOSUB 790  ,810  ,830

1000 IF A$="G" THEN X1=HW/(27.7*PF):GOSUB 850

1010 IF A$="G" THEN 1040

1020 S=F1:W=S*N*D[2*SQR(GF*HW)*FC

1030 GOTO 1050

1040 S=F1:W=S*359*D[2*FC*Y*SQR(PG*HW)
```

```
1050 GOSUB 650

1060 RETURN
```

TANK VOLUME CALCULATOR

How many times have you wondered how much a horizontal tank with F + D heads contains at 15% of full capacity? You do not have to wonder any more. Let TANK tell you in a flash and print a strapping table if you desire. In fact, TANK will compute the volume of any one of the five tanks shown below:

1. Vertical tank with F + D heads
2. Horizontal tank with flat heads
3. Horizontal tank with F + D heads
4. Horizontal tank with hemispherical heads
5. Spherical tank

The program allows you to calculate a specific volume or print a strapping table in desired increments of 100% capacity.

Equations for TANK

Vertical tanks with F + D heads

$$V1 = D^2 * H1 * \pi/4$$
$$V2 = .0809 * D^3 \qquad \text{Head volume}$$
$$V = (V1 + V2 * 2) * 7.45$$

Horizontal tanks with flat heads

$$V1 = L * R^2 * (A1 - SN * CS)$$
$$R = D/2$$
$$CS = 1 - H1/R$$
$$A1 = -ATN [CS/C1 - CS^2)^{.5}] + 1.5703$$
$$SN = (1 - CS^2)^{.5}$$

Horizontal tanks with F + D heads

$$V2 = .2157 * H1^2 (3 * R - H1) \qquad \text{Head volume}$$
$$V = (V1 + V2 * 2) * 7.45$$

Horizontal tanks with hemispherical heads

$$V2 = \pi * H1^2 * (3 * R - H1)/6 \qquad \text{Head volume}$$
$$V = (V1 + V2 * 2) * 7.45$$

Spherical tanks

$V1 = \pi * H1^2 * (3 * R - H1)/3$

$V = V1 * 7.45$

Nomenclature for TANK

VARIABLE	*DESCRIPTION*
L	Tan-to-tan length (ft)
D	Tank diameter (ft)
R	Tank radius (ft)
V	Tank volume (U.S. gal.)
V1	Primary volume (ft³)
V2	Head volume (ft³)
CS	Cosine
SN	Sine
ATN	Arctan function
H1	Tank level (ft)
A1	Temporary variable
π	3.1416

EXAMPLE

RUN "TANK"

TANK VOLUME CALCULATOR

WHAT TYPE OF TANK?

1. VERTICAL TANK/F + D HEADS

2. HORIZONTAL TANK/FLAT HEADS

3. HORIZONTAL TANK/F + D HEADS

4. HORIZONTAL TANK/HEMISPHERICAL HEADS

5. SPHERICAL TANK

TYPE TANK CHOICE(1-5)

TYPE 0 TO END?   **4**

DO YOU WANT TO PRINT A CALIBRATION TABLE? (Y/N)?   **Y**

% INCREMENT(1 = 1%):?   **5**

TYPE TANK DIA(FT), T TO T LENGTH(FT)?   **9,60**

HORIZONTAL TANK/HEMISPHERICAL HEADS

TANK DIA(FT)=9	T TO T LENGTH(FT)=60
% LEVEL	VOLUME-GAL.
5	547.694
10	1555.09
15	2843.03
20	4340.03
25	5999.24
30	7784.77
35	9666.65
40	11618.4
45	13615.8
50	15635.7
55	17655.5
60	19652.9
65	21604.7
70	23486.6
75	25272.1
80	26931.3
85	28428.3
90	29716.2
95	30723.6
100	31275.3

DO YOU WISH TO ANALYZE ANOTHER TANK?(Y/N)?   N

PROGRAM EXECUTION TERMINATED

READY

    LISTING FOR BASIC PROGRAM "TANK"

```
10 CLS
20 PRINT"TANK VOLUME CALCULATOR"
```

```
30 PRINT"WHAT TYPE OF TANK?"
40 PRINT"    1. VERTICAL TANK/ F+D HEADS"
50 PRINT"    2. HORIZONTAL TANK/ FLAT HEADS"
60 PRINT"    3. HORIZONTAL TANK/ F+D HEADS"
70 PRINT"    4. HORIZONTAL TANK/ HEMISPHERICAL HEADS"
80 PRINT"    5. SPHERICAL TANK":PRINT
90 PRINT"TYPE IN TANK CHOICE (1-5)"
100 INPUT"TYPE 0 TO END";C
110 IF C=0 THEN 1560
120 ON C GOTO 130  ,390  ,600  ,810  ,1030
130 REM
140 GOSUB 1250
150 IF A$="Y" THEN 250
160 INPUT"TYPE   TANK DIA(FT),LIQUID DEPTH(FT)";D,H1
170 GOSUB 1290
180 CLS
190 PRINT"VERTICAL TANK/ F+D HEADS"
200 PRINT "DIA(FT)=";D,"LIQUID DEPTH(FT)=";H
210 PRINT"TANK VOLUME(GAL)=";V
220 PRINT"MAX TANK VOLUME(GAL)=";V+V1*2
230 PRINT "VOLUME OF ONE HEAD(GAL)=";V1
240 GOTO 1530
250 INPUT"TYPE   TANK DIA(FT), T TO T LENGTH(FT)";D,H
260 CLS
270 LPRINT"VERTICAL TANK/ F+D HEADS"
280 LPRINT "TANK DIA(FT)=";D,"T TO T LENGTH(FT)=";H
290 LPRINT:LPRINT
300 LPRINT"% LEVEL",,"VOLUME-GAL."
310 FOR I=DX TO 100 STEP DX
```

```
320 H1=D*I/100

330 GOSUB 1290

340 LPRINT I,,V

350 NEXT I

360 PRINT"MAX TANK VOLUME(GAL)=";V+V1*2

370 PRINT"VOLUME OF ONE HEAD(GAL)=";V1

380 GOTO 1530

390 GOSUB 1250

400 IF A$="Y" THEN 480

410 INPUT"TYPE  TANK DIA(FT), LENGTH(FT), LIQUID DEPTH(FT)";D,L,H1

420 GOSUB 1330

430 CLS

440 PRINT"HORIZONTAL TANK/ FLAT HEADS"

450 PRINT"DIA(FT)=";D,"LENGTH(FT)=";L,"FT. LIQ=";H1

460 PRINT"TANK VOLUME(GAL)=";V1

470 GOTO 1530

480 INPUT"TYPE  DIA(FT), TANK LENGTH(FT)";D,L

490 CLS

500 LPRINT"HORIZONTAL TANK/ FLAT HEADS"

510 LPRINT"TANK DIA(FT)=";D,"TANK LENGTH(FT)=";L

520 LPRINT:LPRINT

530 LPRINT"% LEVEL",,"VOLUME-GAL."

540 FOR I=DX TO 100 STEP DX

550 H1=D*I/100

560 GOSUB 1330

570 LPRINT I,,V1

580 NEXT I

590 GOTO 1530
```

```
600 GOSUB 1250

610 IF A$="Y" THEN 690

620 INPUT"TYPE  DIA(FT), T TO T LENGTH(FT), LIQUID DEPTH(FT)";D,L,H1

630 GOSUB 1330

640 CLS

650 PRINT"HORIZONTAL TANK/ F+D HEADS"

660 PRINT"DIA(FT)=";D,"LENGTH(FT)=";L,"FT. LIQ=";H1

670 PRINT "TANK VOLUME(GAL)=";V

680 GOTO 1530

690 INPUT"TYPE  TANK DIA(FT), T TO T LENGTH(FT)";D,L

700 CLS

710 LPRINT"HORIZONTAL TANK/ F+D HEADS"

720 LPRINT"TANK DIA(FT)=";D,"T TO T LENGTH(FT)=";L

730 LPRINT:LPRINT

740 LPRINT"% LEVEL",,"VOLUME-GAL."

750 FOR I=DX TO 100 STEP DX

760 H1=D*I/100

770 GOSUB 1330

780 LPRINT I,,V

790 NEXT I

800 GOTO 1530

810 GOSUB 1250

820 IF A$="Y" THEN 910

830 REM

840 INPUT"TYPE  DIA(FT), T TO T LENGTH(FT),LIQUID DEPTH(FT)";D,L,H1

850 CLS

860 PRINT"HORIZONTAL TANK/ HEMISPHERICAL HEADS"

870 PRINT"DIA(FT)=";D,"LENGTH(FT)=";L,"FT. LIQ=";H1
```

```
880 GOSUB 1330

890 PRINT"TANK VOLUME(GAL)=";V

900 GOTO 1530

910 INPUT"TYPE  TANK DIA(FT),T TO T LENGTH(FT)";D,L

920 CLS

930 LPRINT"HORIZONTAL TANK/ HEMISPHERICAL HEADS"

940 LPRINT"TANK DIA(FT)=";D,"T TO T LENGTH(FT)=";L

950 LPRINT:LPRINT

960 LPRINT"% LEVEL",,"VOLUME-GAL."

970 FOR I=DX TO 100 STEP DX

980 H1=D*I/100

990 GOSUB 1330

1000 LPRINT I,,V

1010 NEXT I

1020 GOTO 1530

1030 GOSUB 1250

1040 IF A$="Y" THEN 1130

1050 REM

1060 INPUT"TYPE  TANK DIA(FT), LIQUID DEPTH(FT)";D,H1

1070 CLS

1080 PRINT"SPHERICAL TANK"

1090 PRINT"TANK DIA(FT)=";D,"LIQUID DEPTH(FT)=";H1

1100 GOSUB 1490

1110 PRINT"TANK VOLUME(GAL)=";V

1120 GOTO 1530

1130 INPUT"TYPE  TANK DIA(FT)";D

1140 CLS

1150 LPRINT "SPHERICAL TANK"
```

```
1160 LPRINT"TANK DIA(FT)=";D

1170 LPRINT:LPRINT

1180 LPRINT"% LEVEL",,"VOLUME-GAL."

1190 FOR I=DX TO 100 STEP DX

1200 H1=D*I/100

1210 GOSUB 1490

1220 LPRINT I,,V

1230 NEXT I

1240 GOTO 1530

1250 INPUT"DO YOU WANT TO PRINT A CALIBRATION TABLE (Y/N)";A$

1260 IF A$="N" THEN 1280

1270 INPUT"% INCREMENT(1=1%):";DX

1280 RETURN

1290 V0=D[2*H1*3.14126*7.45/4

1300 V1=.0809*D[3*7.45

1310 V=V1+V0

1320 RETURN

1330 R=D/2

1340 CS=1-H1/R

1350 IF CS=-1 THEN CS=-.999

1360 A1=-ATN(CS/SQR(1-CS[2))+1.5703

1370 SN=SQR(1-CS[2)

1380 V1=L*R[2*(A1-SN*CS)

1390 IF C=2 THEN V1=V1*7.45:GOTO 1480

1400 IF C=3 THEN 1420

1410 IF C=4 THEN 1450

1420 V2=.21571*H1[2*(3*R-H1)

1430 V=(V1+V2*2)*7.45
```

```
1440 GOTO 1480

1450 V2=3.14126*H1[2*(3*R-H1)/6

1460 V=(V1+V2*2)*7.45

1470 GOTO 1480

1480 RETURN

1490 R=D/2

1500 V1=3.14126*H1[2*(3*R-H1)/3

1510 V=V1*7.45

1520 RETURN

1530 REM-TEST TO CONTINUE

1540 INPUT"DO YOU WISH TO ANALYZE ANOTHER TANK(Y/N)";A$

1550 IF A$="Y" THEN 10

1560 PRINT"PROGRAM EXECUTION TERMINATED"

1570 END
```

DATA ENCRYPTION

Most large companies have equally large distributed computer systems. The potential for theft of sensitive information or trade secrets stored in a computer's memory is very real. When messages are sent via telex from one office to another, it may be necessary to limit the message's distribution to a select group of people. One way to protect information is to encode the data or message in such a way that it will be effectively secure. The program ENCRYPT will take plain text and turn it into a code that will appear useless to unauthorized people. The information is considered effectively secure if it cannot be decoded by unauthorized people before it can do them any good or if it will take too much time and money to decode the information. ENCRYPT is a program that generates an offset code based on the 123 ASCII symbol set. The program will produce code that contains all of the 123 ASCII symbols except the comma, colon, blackslash, carot, underscore, and right bracket. When the main program generates one of these symbols, it is encoded again producing a double or nth offset. This action is similar to the use of a plug board used with the German Engima code machine. More characters can be added to the plug board by adding IF-

THEN statements to the program at lines 1500–1560. The statement should be of the form:

IF B = ASCII CODE THEN 100

The system used by ENCRYPT can be decoded using frequency analysis if enough encoded material is available for sampling. Using the code to send short administrative messages in which the key is changed frequently will make the system secure for all practical purposes. Mixing uppercase and lowercase letters will make the offset even more disguised.

To use the program, enter either plain text or coded text as DATA statements starting at line 2000. The last DATA statement should end the message with "END". The program will prompt you for the key. The key can be any combination of numbers, letters, or symbols, as long as it is less than 123 characters long. The program will then encode or decode the message. It is easy to see that unless some unauthorized person has a working knowledge of cryptology, the information contained in the examples would be totally useless.

REFERENCE

Roberts, R., Byte, April 1982, pg. 432.

EXAMPLE

RUN

ENCODING/DECODING PROGRAM/

TYPE IN KEY:? **MARY HAD A LITTLE LAMB**

KEY = MARY HAD A LITTLE LAMB

ENCODING(1) OR DECODING(2):? **1**

ANTICIPATE MERGER IN THE NEXT FEW DAYS.

(5;0*07(;nb4n9.n9b05b;/nb5n?;b − n > b + (@!p

BUY STOCK AND OPTIONS NOW.

+ > Bd < = 8p4d*7 − d89 = 287 < d78@r

REPLY BY TELEX USING KEY #2.

>1<8Eg.Eg@181DgA?5&3g71Egjyu

SIGNED

D+8?65

DENNIS WRIGHT-THE BROKERAGE HOUSE

0;DD?IqMH?=>J#J>;q8HEA;H7=;q>EKI;

LIST COMPLETE MESSAGE TO PRINTER(Y/N)?   **Y**

(5;0*07(;nb4n9.n9b05b;/nb5n?;b−n>b+(@!p

+>Bd<=8p4d*7−d89=287<d78@r

>1<8Eg.Eg@181DgA?5&3g71Egjyu

D+8?65

0;DD?IqMH?=>J#J>;q8HEA;H7=;q>EKI;

PROGRAM EXECUTION TERMINATED

READY

>

```
    LISTING FOR BASIC PROGRAM "ENCRYPT"


10 CLEAR 5000:DIM B$(50),D$(50)

20 REM-ADAPTED FROM ROBERTS,R.,BYTE,MC GRAW HILL, NY,NY,pp.432,APRIL,
   1982 WITH PERMISSION MC GRAW HILL @ 1982

30 CLS:PRINT TAB(15);"ENCODING/DECODING PROGRAM":LPRINT

40 INPUT"TYPE  KEY:";P$:GOTO 240

42 LPRINT

45 LPRINT"KEY=";P$

50 INPUT"ENCODING(1) OR DECODING(2):";I

55 LPRINT

60 IF I=2 THEN A=-A
```

```
65 N=1

70 READ B$

72 D$=""

75 IF B$="END" THEN 210

80 FOR X=1 TO LEN(B$)

90 B=ASC(MID$(B$,X,1))

100 IF I=1 THEN B=B+A:IF B>123 THEN B=(B-123)+32

110 IF I=2 THEN B=B+A:IF B<32 THEN B=(B+123)-32

112 GOTO 1500

120 D$=D$+CHR$(B)

130 NEXT X

140 IF I=1 THEN A=A+INT(ABS(5*SIN(3.1416*N/10)))+1:IF A>=124 THEN A=1

150 IF I=2 THEN A=A-INT(ABS(5*SIN(3.1416*N/10)))-1:IF A<=-124 THEN A=-
       1

160 B$(N)=B$

170 D$(N)=D$

180 LPRINT B$:LPRINT D$:LPRINT

190 N=N+1

200 GOTO 70

210 GOSUB 1000:PRINT"PROGRAM EXECUTION TERMINATED"

212 END

240 P=LEN(P$):FOR X=1 TO LEN(P$)

245 TEMP$=MID$(P$,X,1)

250 P1=P1+ASC(TEMP$):NEXT X

260 A=INT(P1/P):IF A>123 THEN PRINT"KEY TOO LARGE....TRY AGAIN":GOTO 2
       0

270 GOTO 42

1000 CLS

1010 INPUT"LIST COMPLETE MESSAGE TO PRINTER(Y/N)";A$

1020 IF A$="N" THEN RETURN
```

```
1030 FOR I=1 TO N-1

1040 LPRINT D$(I)

1050 LPRINT

1060 NEXT I

1070 RETURN

1500 IF B=44 THEN GOTO 100

1510 IF B=58 THEN GOTO 100

1520 IF B=34 THEN GOTO 100

1530 IF B=92 THEN GOTO 100

1540 IF B=93 THEN GOTO 100

1550 IF B=94 THEN GOTO 100

1560 IF B=95 THEN GOTO 100

1570 GOTO 120

2000 DATA ANTICIPATE MERGER IN THE NEXT FEW DAYS.

2010 DATA BUY STOCK AND OPTIONS NOW.

2020 DATA REPLY BY TELEX USING KEY #2.

2030 DATA SIGNED

2040 DATA DENNIS WRIGHT-THE BROKERAGE HOUSE

2050 DATA END
```

ECONOMICAL INSULATION THICKNESS

INSUL is a program that will determine the most economical thickness of insulation for hot pipes. The program calculates the bare pipe heat loss and then performs an incremental discounted cash flow analysis for each thickness of insulation applied to the pipe as specified by you. You must supply the following data:

1. Pipe OD (in)
2. Pipe temperature (°F)
3. Bare pipe emissivity
4. Insulation cover emissivity (Table 8.1)
5. Thermal conductivity of the insulation (see Figure 8.1)

Table 8.1. Typical Emissivity Values[2]

MATERIAL	TEMPERATURE RANGE (°F)	EMISSIVITY
Aluminum		
Polished	440–1070	.038–.057
Oxidized	390–1110	.110–.190
Copper		
Polished	242	.023
Heat treated at 1100	390–1100	.570
Cast iron		
Heat treated at 1100	390–1100	.640–.780
Steel		
Oxidized at 1100	390–1100	.790
Nickel		
Polished	74	.045
Paints		
Black lacquer	76	.875
White lacquer	100–200	.80–.958
Oil paints (16 colors)	212	.920–.960
10% Aluminum 22% lacquer	212	.520
26% Aluminum 27% lacquer	212	.300
Primer	66	.924

6. Outside air temperature (°F)
7. The cost to apply various depths of insulation ($/ft)
8. The cost of energy ($/mm Btu)
9. The discount factor for investments (Decimal)

The program calculates the following:

1. Bare pipe heat loss
2. Insulated pipe heat loss for each depth of insulation
3. Outside temperature of the pipe
4. Profitability index for each incremental layer of insulation

In general, enough insulation should be added to reduce the insulated pipe heat loss to 5% of the bare pipe heat loss or until the profitability index for that layer of insulation is close to 1.00. Cost data for insulation are available from Richardson's[1] or from a local contractor. Estimated installed costs for calcium silicate insulation are listed in Appendix A.

---

[1]*Process Plant Construction Estimating Standards,* Richardson Engineering Services, Inc., San Marcos, California, 1984, pg. *15*-80., vol. III.

[2]Excerpted by special permission from *Chemical Engineering* (Jan. 25) Copyright © 1982 by McGraw-Hill, New York, NY 10020.

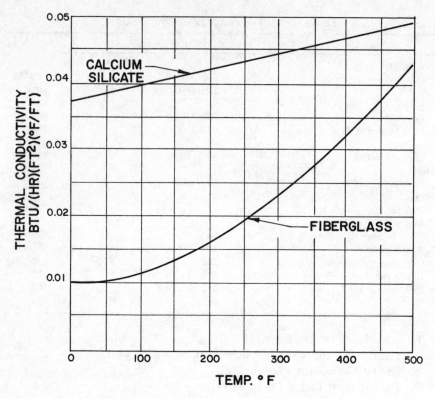

Figure 8.1. Insulation Thermal Conductivities

Equations for INSUL

$Q_1 = 2 * \pi * K1 \, (T1 - T2)/\ln(D2/D1)$

$Q_2 = .27 \, [(T2 - T3) * 12/D2]^{.25} * [\pi * D2/12 * (T2 - T3)]$

$Q_3 = .173 * E * \pi * D2/12 * \{[(T2 + 460)/100]^4 - [(T3 + 460)/100]^4\}$

$Q_4 = Q_2 + Q_3$

$C1_T = Q_1 * .0084 * E3$

$DCF = \sum_{N=1}^{5} (C1_T - C1_{T-1}) * (1 - TXR) * [(1 + N1)^N / (1 + N2)^N]$

$PI = DCF/(C_T - C_{T-1})$

Nomenclature for INSUL

VARIABLE	DESCRIPTION
Q_1	Conduction heat loss (Btu/hr-ft)
Q_2	Convection heat loss (Btu/hr-ft)

Q_3	Radiation heat loss (Btu/hr-ft)
Q_4	Total heat loss (Btu/hr-ft)
T1	Pipe fluid temperature (°F)
T2	Surface temperature of insulated pipe (°F)
T3	Air temperature (°F)
Kl	Thermal conductivity of insulation (Btu · in./hr ft² °F)
E1	Pipe emissivity
E2	Insulation cover emissivity
E3	Cost of energy ($/Million Btu)
Cl_T	Annual cost of heat loss ($/yr.-ft)
DCF	Total discounted cash flow ($-5 yr.)
TXR	Tax rate (decimal)
PI	Profitability index
D1	Pipe OD (in.)
D2	Pipe and insulation OD (in.)
C_T	Total cost of each insulation thickness ($/ft)
N1	Inflation rate (Decimal)
N2	Discount rate (Decimal)
π	3.1416

EXAMPLE

RUN

ECONOMICAL INSULATION THICKNESS FOR HOT PIPES

FUEL INFLATION RATE(DEC) = ?   **.10**

DISCOUNT RATE(DEC) = ?   **.15**

TAX RATE(DEC) = ?   **.46**

NO. OF INSULATION LAYERS = ?   **3**

PIPE OD (IN) = ?   **6.5**

PIPE TEMP (F) = ?   **350**

AMBIENT TEMP (F) = ?   **60**

INSULATION K = ?   **.043**

BARE PIPE EMISSIVITY = ?   **.75**

INSULATION COVER EMISSIVITY = ?   **.10**

COST OF ENERGY ($/MM BTU)=? **10.50**

ENTER THICKNESS OF LAYER NO.1=? **1**

COST ($/FT) FOR LAYER NO.1=? **13**

ENTER THICKNESS OF LAYER NO.2=? **2**

COST ($/FT) FOR LAYER NO. 2=? **21**

ENTER THICKNESS OF LAYER NO. 3=? **3**

COST ($/FT) FOR LAYER NO. 3=? **31**

INSULATION K=.043

PIPE OD (IN)=6.5   BARE PIPE TEMP (F)=350

BARE PIPE HEAT LOSS(BTU/HR-FT)=1429.79

PIPE OD.	HEAT LOSS	INSULATION	OD	PI RATIO
INCHES	BTU/HR-FT	INCHES	TEMP(F)	INDEX
8.5	202.776	1	148.637	19.7098
10.5	132.339	2	115.067	1.83861
12.5	103.16	3	100.288	.609317

PROGRAM EXECUTION TERMINATED

READY

    LISTING FOR BASIC PROGRAM "INSUL"

```
10 DIM C(5),C1(5):CLS

20 PRINT "ECONOMICAL INSULATION THICKNESS FOR HOT PIPES"

30 PRINT

40 INPUT"FUEL INFLATION RATE(DEC)=";N1

50 INPUT"DISCOUNT RATE(DEC)=";N2
```

```
 60 INPUT"TAX RATE(DEC)=";TXR

 70 INPUT"NO. OF INSULATION LAYERS=";Z1

 80 INPUT"PIPE OD(IN)=";D1

 90 INPUT"PIPE TEMP(F)=";T1

100 INPUT"AMBIENT TEMP(F)=";T3

110 INPUT"INSULATION K=";K1

120 INPUT"BARE PIPE EMISSIVITY=";E1

130 INPUT"INSULATION COVER EMISSIVITY=";E2

140 INPUT"COST OF ENERGY($/MM BTU)=";E3

150 CLS

160 FOR I=1 TO Z1

170 PRINT"ENTER THICKNESS OF LAYER NO.";I;"=";

180 INPUT Z2(I)

190 PRINT"COST ($/FT) FOR LAYER NO.";I;"=";

200 INPUT C(I)

210 CLS

220 NEXT I

230 TZ=T1

240 CLS

250 GOSUB 560

260 PRINT  : PRINT

270 PRINT "==============================================================
    ="

280 PRINT "INSULATION K=";K1

290 PRINT "PIPE OD(IN)=";D1;TAB(25);"BARE PIPE TEMP(F)=";T1

300 PRINT "==============================================================
    ="

310 PRINT "BARE PIPE HEAT LOSS(BTU/HR-FT)=";Q6

320 DZ=D1
```

```
330 PRINT "PIPE OD.";TAB(15);"HEAT LOSS";TAB(30);"INSULATION";TAB(45);
    "OD";TAB(54);"PI RATIO"

340 PRINT "INCHES";TAB(15);"BTU/HR-FT";TAB(30);"INCHES";TAB(45);"TEMP(
    F)";TAB(54);"INDEX"

350 FOR I1=1 TO Z1

360 D2=D1+2*Z2(I1)

370 T2=T3+10:TEMP=T2

380 GOSUB 650

390 F1=F

400 T2=T2+.001

410 GOSUB 650   :T2=T2-.001

420 DERV=(F-F1)/.001

430 T2=T2-F1/DERV

440 IF ABS(T2-TEMP)<=.01 THEN 460

450 TEMP=T2:GOTO 380

460 C1(I1)=Q1*.0084*E3

470 T=0

480 FOR I2=1 TO 5

490 T=T+(C1(I1-1)-C1(I1))*(1-TXR)*(1+N1)[I2/(1+N2)[I2

500 NEXT I2

510 P1=T/(C(I1)-C(I1-1))

520 PRINT D2;TAB(15);Q1;TAB(30);(D2-D1)/2;TAB(45);T2;TAB(54);P1

530 NEXT I1

540 PRINT "PROGRAM EXECUTION TERMINATED"

550 GOTO 750

560 A1=((T2-T3)*12/D1)[.25

570 A2=3.14126*D1/12*(T2-T3)

580 Q4=.27*A1*A2

590 A3=.173*E1*3.14126*D1/12
```

```
600 A4=((T2+460)/100)[4-((T3+460)/100)[4

610 Q5=A3*A4

620 Q6=Q4+Q5

630 C1(0)=Q6*.0084*E3

640 RETURN

650 Q1=2*3.14126*K1*(T1-T2)/LOG(D2/D1)

660 A1=((T2-T3)*12/D2)[.25

670 A2=3.14126*D2/12*(T2-T3)

680 Q2=.27*A1*A2

690 A3=.173*E2*3.14126*D2/12

700 A4=((T2+460)/100)[4-((T3+460)/100)[4

710 Q3=A3*A4

720 Q4=Q2+Q3

730 F=Q1-Q4

740 RETURN

750 END
```

APPENDIX

Table A.1. Vapor Pressures of Organic Compounds:
$\log_{10}P = (-0.2185\,A/K) + B$

FORMULA	NAME	A	B	RANGE
CBrN	Cyanogen bromide	10882.8	10.03204	−35.7 to 61.5
CBrF$_3$	Bromotrifluoromethane	—	—	—
CClF$_3$	Chlorotrifluoromethane	3996.3	7.31428	−149.5 to 52.8
CCl$_2$F$_2$	Dichlorodifluoromethane	8363.1	9.33487	−76.7 to 13.1
CCl$_3$F	Trichlorofluoromethane	6424.1	7.56378	−84.3 to 194.0
CCl$_4$	Carbon tetrachloride	9271.5	8.05000	−70.0 to −50.0
		7628.8	7.58604	−50.0 to 276.0
CF$_4$	Carbon tetrafluoride	3016.5	7.38060	−184.6 to −127.7
CHClF$_2$	Chlorodifluoromethane	5212.9	7.73004	−122.8 to 85.3
CHCl$_3$	Chloroform	7500.5	7.73508	−58.0 to 254.0
CHN	Hydrogen cyanide	7338.8	8.16845	−70.8 to 183.5
CH$_2$Cl$_2$	Dichloromethane	7572.3	8.18330	−70.0 to 40.7
CH$_2$O$_2$	Formic acid	9896.5	8.77933	−20.0 to 100.6
CH$_3$BO	Borine carbonyl	4867.6	7.97636	−96.3 to 190.0
CH$_3$Cl	Methyl chloride	5373.3	7.54620	−99.5 to 137.5
CH$_3$F	Methyl fluoride	3986.4	7.34492	−147.3 to 43.5
CH$_3$NO	Formamide	15556.6	9.88740	70.5 to 210.5
CH$_4$	Methane	2128.8	7.02772	−205.9 to −86.3
CH$_4$O	Methanol	9377.2	8.95470	−62.0 to −44.0
		8978.8	8.63982	−44.0 to 224.0
CH$_5$ClSi	Chloromethylsilane	6349.5	7.84384	−95.0 to 8.7
CH$_6$Si	Methylsilane	4683.6	7.66043	−138.5 to −56.9
CIN	Cyanogen iodide	14065.4	10.29357	25.2 to 141.1
CO	Carbon monoxide	1613.3	7.10056	222.0 to −141.9
COSe	Carbonyl selenide	5366.5	7.55290	−117.1 to 21.9
CSSe	Carbon selenosulfide	8003.0	7.78588	−47.3 to 85.6
C$_2$BrCl$_2$O	Trichloroacetyl bromide	9673.9	7.98533	−7.4 to 143.0
C$_2$ClF$_3$	1-Chloro-1,2,2-trifluoroethylene	5421.5	7.67920	−116.0 to 91.9

(continued)

313

Vapor Pressures of Organic Compounds (continued)

FORMULA	NAME	A	B	RANGE
C_2F_4	Tetrafluoroethylene	—	—	—
$C_2Cl_2F_4$	1,2-Dichloro-1,1,2,2-tetrafluoroethane	6134.6	7.67560	−95.4 to 140.9
C_2Cl_4	Tetrachloroethylene	9240.5	8.02620	−20.6 to 120.8
$C_2Cl_4F_2$	1,1,2,2-Tetrachloro-1,2-difluoroethane	8746.2	8.12294	−37.5 to 92.0
C_2Cl_6	Hexachloroethane	11711.3	8.44062	32.7 to 185.6
C_2F_6	Hexafluoroethane	—	—	—
C_2HCl_3	Trichloroethylene	8314.7	7.95634	−43.8 to 86.7
$C_2HCl_3O_2$	Trichloroacetic acid	13817.9	9.34143	51.0 to 195.6
C_2H_2	Acetylene	4665.8	8.08484	−142.9 to 34.8
$C_2H_2Br_4$	1,1,2,2-Tetrabromoethane	12911.5	8.35719	65.0 to 243.5
$C_2H_2Cl_2$	cis-1,2-Dichloroethylene	7420.6	7.68513	−58.4 to 260.0
$C_2H_2Cl_2$	1,1-Dichloroethylene	7211.8	8.08714	−77.2 to 31.7
$C_2H_2Cl_2O_2$	Dichloroacetic acid	12952.9	8.94660	44.0 to 194.4
$C_2H_2Cl_4$	1,1,2,2-Tetrachloroethane	9917.1	8.07258	−3.8 to 145.9
$C_2H_3BrO_2$	Bromoacetic acid	13537.8	9.04192	54.7 to 208.0
C_2H_3Cl	1-Chloroethylene	6263.0	8.20278	−105.1 to −13.8
$C_2H_3Cl_3$	1,1,1-Trichloroethane	8012.7	7.95590	−52.0 to 74.1
$C_2H_3ClF_2$	1-Chloro-1,1-difluoroethane	—	—	—
$C_2H_3Cl_3O_2$	Trichloroacetaldehyde hydrate	12141.5	10.07393	−9.8 to 96.2
C_2H_3N	Acetonitrile	8173.2	7.93866	−47.0 to 81.8
C_2H_3NS	Methyl isothiocyanate	7990.1	7.27281	−34.7 to 119.0
C_2H_4BrCl	1-Bromo-1-chloroethane	9314.9	8.70076	−36.0 to 82.7
$C_2H_4Br_2$	1,2-Dibromoethane	9229.4	7.93538	−27.0 to 304.5
$C_2H_4Cl_2$	1,2-Dichloroethane	7950.7	7.70840	−44.5 to 285.0
C_2H_4O	Acetaldehyde	7267.8	8.32780	−81.5 to 20.2
$C_2H_4O_2$	Acetic acid	9963.9	8.50200	−35.0 to 10.0
		9486.6	8.14240	−17.2 to 312.5
$C_2H_4O_2S$	Mercaptoacetic acid	13790.7	9.04519	60.0 to 154.0
C_2H_5Cl	Ethyl chloride	6310.6	7.66020	−89.8 to 180.5
$C_2H_5Cl_3Si$	Trichloroethylsilane	9457.8	8.45568	−27.9 to 99.5
C_2H_5F	Ethyl fluoride	5519.5	7.86070	−117.0 to 90.0
C_2H_5I	Ethyl iodide	7851.8	7.88096	−54.4 to 72.4
C_2H_5NO	Acetaldoxime	11317.8	9.27297	−5.8 to 115.0
$C_2H_5N_3O_2$	Di(nitrosomethyl)amine	10326.7	8.19428	3.2 to 153.0
$C_2H_6Cl_2Si$	Dichlorodimethylsilane	7995.7	7.99214	−53.5 to 70.3
C_2H_6O	Dimethyl ether	5409.8	7.58547	−115.7 to 125.2
C_2H_6S	Dimethyl sulfide	6742.3	7.58920	−75.6 to 224.5
C_2H_6Sb	Dimethylantimony	12075.7	8.33544	44.0 to 211.0
C_2H_7N	Dimethylamine	6660.0	7.99516	−87.7 to 162.6
C_2H_8Si	Dimethylsilane	5497.8	7.66117	−115.0 to −20.1
$C_2H_{11}NSi_2$	2-Ethyldisilazane	7348.3	7.64762	−62.0 to 65.9
C_3N_3N	Acrylonitrile	7941.4	7.85101	−51.0 to 78.5
C_3H_4	Propyne	5632.4	7.75141	−111.0 to 125.0
$C_3H_4Cl_2O_2$	Methyl dichoroacetate	10820.5	8.58442	3.2 to 143.0
$C_3H_4O_2$	Acrylic acid	10955.1	8.65970	3.5 to 141.0

Vapor Pressures of Organic Compounds (continued)

FORMULA	NAME	A	B	RANGE
$C_3H_5Br_3$	1,2,3-Tribromopropane	12047.1	8.23707	47.5 to 220.0
C_3H_5Cl	Allyl chloride	7386.8	7.99195	−70.0 to 44.6
$C_3H_5ClO_2$	Methyl chloroacetate	10815.0	8.78863	−2.9 to 130.3
$C_3H_5Cl_3$	1,2,3-Trichloropropane	10714.3	8.32689	9.0 to 158.0
C_3H_5N	Propionitrile	8769.0	8.07947	−35.0 to 97.1
C_3H_5NS	Ethylisothiocyanate	9574.7	8.06147	−13.2 to 131.0
C_3H_6	Propene	4697.4	7.39391	−131.9 to 85.0
C_3H_6BrNO	2-Bromo-2-nitrosopropane	9619.6	8.80638	−33.5 to 83.0
C_3H_6Br	1,3-Dibromopropane	1037.4	8.04508	9.7 to 141.6
$C_3H_6Cl_2$	1,2-Dichloropropane	8428.5	7.88241	−38.5 to 96.8
C_3H_6O	Acetone	7641.5	7.90402	−59.4 to 214.5
C_3H_6O	Propylene oxide	7295.8	8.09347	−75.0 to 34.5
CiH_6O_2	Methyl acetate	7732.8	7.93878	−57.2 to 225.0
$C_3H_6O_3$	Methyl glycolate	11105.0	8.60777	9.6 to 151.5
C_3H_7Br	n-Propyl bromide	8029.8	8.01122	−53.0 to 71.0
C_3H_7Cl	n-Propyl chloride	7485.7	8.02873	−68.3 to 46.4
		6905.8	7.59300	.0 to 50.0
$C_3H_7Cl_3Si$	Trichloroisopropylsilane	8973.3	7.91444	−24.3 to 118.5
C_3H_7I	2-Iodopropane	8243.4	7.87382	−43.3 to 89.5
$C_3H_7NO_2$	1-Nitropropane	9949.9	8.27862	−9.6 to 131.6
$C_3H_7NO_2$	Ethyl carbamate	13078.6	9.14254	65.8 to 184.0
C_3H_8O	n-Propanol	11298.8	9.51800	−45.0 to −10.0
		10421.1	8.93729	−15.0 to 250.0
C_3H_8O	Ethyl methyl ether	6388.3	7.80575	−91.0 to 160.0
$C_3H_8O_2$	1,3-Propanediol	13782.3	9.05397	59.4 to 214.2
$C_3H_8O_2$	Glycerol	18188.9	10.01579	125.5 to 290.0
C_3H_8S	Propanethiol	7855.3	7.94089	−56.0 to 67.4
C_3H_9ClSi	Chlorotrimethylsilane	7589.1	7.92044	−62.8 to 57.9
C_3H_9N	n-Propylamine	7408.0	7.86799	−64.4 to 214.5
$C_3H_9O_4P$	Trimethyl phosphate	11019.7	8.05848	26.0 to 192.7
C_3O_2	Carbon suboxide	6446.3	7.94818	−94.8 to 6.3
$C_4Cl_6O_3$	Trichloroacetic anhydride	12929.0	8.58314	56.2 to 223.0
C_4H_4O	Furan	—	—	
$C_4H_2Br_2O_3$	α,β-Dibromomaleic anhydride	12579.2	8.52395	50.0 to 215.0
$C_4H_2O_3$	Maleic anhydride	12122.3	8.47641	44.0 to 202.0
C_4H_4	Butenyne	6677.2	8.16150	−93.2 to 5.3
$C_4H_4Cl_2O_3$	Chloroacetic anhydride	14645.1	9.39589	67.2 to 217.0
$C_4H_4O_4$	1,4-Dioxane-2,6-dione	14013.6	8.85568	103.0 to 240.0
C_4H_4Se	Selenophene	7766.1	7.29068	−39.0 to 114.3
$C_4H_5ClO_3$	Ethyl chloroglyoxylate	10268.4	8.40053	−5.1 to 135.0
C_4H_5N	3-Butenenitrile	9447.8	8.17542	−19.6 to 119.0
C_4H_5N	cis-Crotononitrile	8905.4	8.00622	−29.0 to 108.0
$C_4H_5NO_2$	Succinimide	16422.0	9.30318	115.0 to 287.5
C_4H_6	1,2-Butadiene	6539.1	7.81848	−89.0 to 18.5
C_4H_6	Cyclobutene	6167.5	7.79548	−99.1 to 2.4
C_4H_6	2-Butyne	7868.5	8.70592	−73.0 to 27.2

(continued)

Vapor Pressures of Organic Compounds (continued)

FORMULA	NAME	A	B	RANGE
$C_4H_6Cl_2O_2$	2-Chloroethyl chloroacetate	12588.7	8.65356	46.0 to 205.0
$C_4H_6O_2$	trans-Crotonic acid	13252.2	9.20563	80.0 to 185.0
$C_4H_6O_2$	Methacrylic acid	12526.6	9.19944	25.5 to 161.0
$C_4H_6O_2$	Acetic anhydride	10930.4	8.70800	1.7 to 288.5
C_4H_7Br	cis-1-Bromo-1-butene	8300.2	7.95302	−44.0 to 86.2
C_4H_7Br	2-Bromo-1-butene	8389.7	8.10530	−43.7 to 81.0
C_4H_7Br	trans-2-Bromo-2-butene	8238.1	7.92759	−45.0 to 85.0
C_4H_7BrO	2-Methylpropionyl bromide	10974.6	8.39340	13.5 to 163.0
$C_4H_7Br_3$	1,2,2-Tribromobutane	11622.3	8.11636	41.0 to 213.8
$C_4H_7ClO_2$	Ethyl chloroacetate	10522.6	8.39747	1.0 to 144.2
$C_4H_7Cl_3$	1,2,3-Trichlorobutane	9447.0	7.57526	0.5 to 169.0
$C_4H_7NO_2$	Diacetamide	14508.1	9.29208	70.0 to 223.0
C_4H_8	cis-2-Butene	6401.0	7.97658	−96.4 to 3.7
C_4H_8	2-Methylpropene	5742.9	7.60156	−105.1 to −6.9
C_4H_8BrClO	2-Bromethyl-2-chloroethyl ether	12010.5	8.50063	36.5 to 195.8
$C_4H_8Br_2$	dl-2,3-Dibromobutane	10136.1	8.00377	5.0 to 160.5
$C_4H_8Br_2$	1,4-Dibromobutane	11369.3	8.17360	32.0 to 197.5
$C_4H_8Cl_2$	1,2-Dichlorobutane	8850.6	7.78627	−23.6 to 123.5
$C_4H_8Cl_2$	1,1-Dichloro-2-methylpropane	8765.6	7.96629	−31.0 to 105.0
$C_4H_8Cl_2$	1,3-Dichloro-2-methylpropane	10519.7	8.52338	−3.0 to 135.0
C_4H_8O	1,2-Epoxy-2-methylpropane	7066.6	7.61384	−69.0 to 55.0
$C_4H_8O_2$	Dioxane	8546.2	7.86411	−35.8 to 101.1
$C_4H_8O_2$	Isobutyric acid	11182.8	8.55028	14.7 to 336.0
$C_4H_8O_2$	Methyl propanoate	8356.2	7.99898	−42.0 to 239.0
$C_4H_8O_2$	Isopropyl formate	8230.2	8.17007	−52.0 to 68.3
$C_4H_8O_3$	Ethyl glycolate	11318.1	8.62684	14.3 to 158.2
C_4H_9BrO	1-Bromo-2-butanol	13473.7	9.93967	23.7 to 145.0
C_4H_9Cl	sec-Butyl chloride	7407.9	7.62562	−60.2 to 68.0
C_4H_9Cl	tert-Butyl chloride	6876.0	7.51712	−19.0 to 51.0
C_4H_9I	n-Butyl iodide	—	—	—
$C_4H_9NO_2$	Ethyl methylcarbamate	12161.2	8.89205	26.5 to 170.0
$C_4H_9N_2O_2$	Di(nitrosoethyl)amine	10894.8	8.18818	18.5 to 176.9
C_4H_{10}	2-Methylpropane	5084.4	7.25000	−115.0 to −34.0
$C_4H_{10}Cl_2Si$	Dichlorodiethylsilane	10038.6	8.34194	−9.2 to 130.4
$C_4H_{10}O$	n-Butyl alcohol	10970.5	8.92959	−1.2 to 277.0
$C_4H_{10}O$	Isobutyl alcohol	10936.0	9.11803	−9.0 to 251.0
$C_4H_{10}O$	Diethyl ether	6946.2	7.75665	−74.3 to 183.3
$C_4H_{10}O_2$	1,3-Butanediol	10479.1	7.53193	22.2 to 206.5
$C_4H_{10}O_2$	1,2-Dimethoxyethane	7681.0	7.49342	−48.0 to 93.0
$C_4H_{10}O_3$	Diethylene glycol	16146.7	9.66817	91.8 to 244.8
$C_4H_{10}O_3S$	Diethyl sulfite	10783.0	8.36122	10.0 to 159.0
$C_4H_{10}S$	n-Butanethiol	—	—	—
$C_4H_{10}Se$	Diethyl selenide	9274.7	8.22595	−25.7 to 108.0
$C_4H_{11}N$	Diethyl amine	7307.5	7.70171	−33.0 to 210.0
$C_4H_{12}Cl_2Si_2$	1,3-Dichlorotetramethyldisiloxane	9881.6	8.15490	−7.4 to 138.0

Vapor Pressures of Organic Compounds (continued)

FORMULA	NAME	A	B	RANGE
$C_4H_{12}Si$	Tetramethylsilane	6439.2	7.53767	-83.8 to 178.0
$C_4H_{14}B_2$	Tetramethyldiborane	7517.1	7.68899	-59.6 to 68.6
C_4F_{10}	Perfluoro-n-butane	—	—	—
C_5H_4ClN	2-Chloropyridine	10614.5	8.12737	13.3 to 170.2
$C_5H_4O_3$	Citraconic anhydride	12307.8	8.42188	47.1 to 213.5
$C_5H_6Cl_2O_2$	Glutaryl chloride	3192.1	8.76924	56.1 to 217.0
$C_5H_6O_2$	Furfuryl alcohol	12815.8	9.19802	31.8 to 170.0
$C_5H_6O_3$	Pyrotaraaric anhydride	13251.2	8.46903	69.7 to 247.4
C_5H_6S	3-Methylthiophene	9084.1	8.02060	-24.5 to 115.4
C_5H_7N	Tiglonitrile	8704.6	7.72132	-25.5 to 122.0
C_5H_8N	α-Ethylacrylonitrile	8679.1	7.80407	-29.0 to 114.0
C_5H_8	Cyclopentene	—	—	—
C_5H_8	1,3-Pentadiene	7313.9	7.98323	-71.8 to 42.1
C_5H_8O	Tiglaldehyde	9009.2	7.94452	-25.0 to 116.4
$C_5H_8O_2$	Tiglic acid	13756.5	9.26320	52.0 to 198.5
$C_5H_8O_2$	α-Ethylacrylic acid	14417.8	9.85865	47.0 to 179.2
$C_5H_8O_2$	Methyl methacrylate	8974.9	8.14094	-30.5 to 101.0
$C_5H_8O_4$	Glutaric acid	22085.2	11.26789	155.5 to 303.0
$C_5H_9ClO_2$	Ethyl α-chloropropionate	11032.8	8.64474	6.6 to 146.5
C_5H_9N	Valeronitrile	9931.3	8.14864	-6.0 to 140.8
C_5H_{10}	1-Pentene	6931.2	7.91496	-80.4 to 30.1
C_5H_{10}	3-Methyl-2-butene	7112.8	7.88971	-75.4 to 38.5
C_5H_{10}	3-Methyl-1-butene	—	—	—
C_5H_{10}	Methylcyclobutane	7413.2	7.97240	-96.0 to 4.5
$C_5H_{10}Br_2$	1,2-Dibromo-2-methylbutane	7616.9	6.85371	-28.8 to 149.0
$C_5H_{10}Cl_2Si$	Allyldichloroethylsilane	9833.9	7.98265	-3.0 to 150.3
$C_5H_{10}O$	Methyl n-propyl ketone	11240.6	9.43208	-12.0 to 103.3
$C_5H_{10}Cl_2O$	2-Chloroethyl 2-chloroisopropyl ether	11420.8	8.41126	24.7 to 180.0
$C_5H_{10}Cl_2O_2$	Di (2-chloroethoxy)methane	12908.0	8.66854	53.0 to 185.0
$C_5H_{10}O_2$	Valeric acid	13370.3	9.27117	42.2 to 184.4
$C_5H_{10}O_2$	Ethyl propanoate	8877.8	8.03304	-28.0 to 264.5
$C_5H_{10}O_2$	Isopropyl acetate	8794.8	8.21226	-38.3 to 89.0
$C_5H_{10}O_2$	Methyl isobutyrate	8593.3	7.95875	-34.1 to 259.5
$C_5H_{10}O_2$	Isobutyl formate	8678.8	7.96193	-32.7 to 261.0
$C_5H_{10}O_2$	tert-Butyl formate	8955.3	8.18137	-32.7 to 98.0
$C_5H_{11}Br$	1-Bromopentane	—	—	—
$C_5H_{11}Br$	1-Chloropentane	—	—	—
$C_5H_{11}I$	1-Iodo-3-methylbutane	9951.6	8.06638	-2.5 to 148.2
$C_5H_9O_2$	Pentanoic acid	—	—	—
$C_5H_{11}NO_3$	Isoamyl nitrate	10817.2	8.52434	5.2 to 147.5
C_5H_{12}	2-Methylbutane	6470.8	7.54468	-82.9 to 180.3
$C_5H_{12}O$	Amyl alcohol	12495.5	9.57434	13.6 to 137.8
$C_5H_{12}O$	2-Pentanol	12086.2	9.64871	1.5 to 119.7
$C_5H_{12}O$	Ethyl propyl ether	7092.7	7.50325	-64.3 to 223.0
$C_5H_{12}O_3$	2,3,4-Pentanetriol	19694.4	10.01839	155.0 to 327.2

(continued)

Vapor Pressures of Organic Compounds (continued)

FORMULA	NAME	A	B	RANGE
$C_5H_{14}OSi$	Ethoxytrimethylsilane	8030.6	7.93511	−50.9 to 75.7
$C_5H_{14}Sn$	Ethyltrimethyltin	8820.9	7.95247	−30.0 to 108.8
C_6Cl_6	Hexachlorobenzene	15199.1	8.55049	114.4 to 309.4
C_6HCl_5O	Pentachlorophenol	16742.6	9.15020	192.2 to 309.3
$C_6H_2Cl_4$	1,2,3,4-Tetrachlorobenzene	12872.5	8.25105	68.5 to 254.0
$C_6H_2Cl_4$	1,2,4,5-Tetrachlorobenzene	128.28.8	8.28221	146.0 to 245.0
$C_6H_3BrCl_2O$	2-Bromo-4,6-dichlorophenol	13829.1	8.47404	84.0 to 268.0
$C_6H_3Cl_3$	1,2,4-Trichlorobenzene	11425.1	8.03052	38.4 to 213.0
$C_6H_3Cl_3O$	2,4,5-Trichlorophenol	13237.0	8.40107	72.0 to 251.8
$C_6H_4Br_2$	1,4-Dibromobenzene	13047.8	8.76977	61.0 to 218.6
$C_6H_4Cl_2$	1,2-Dichlorobenzene	10943.0	8.18527	20.0 to 179.0
$C_6H_4Cl_2$	1,4-Dichlorobenzene	17260.5	12.48000	30.0 to 50.0
		10611.0	8.07363	58.8 to 173.9
$C_6H_4Cl_2O$	2,6-Dichlorophenol	13472.0	8.86400	59.5 to 220.0
$C_6H_5AsCl_2$	Dichlorophenylarsine	12229.5	7.90607	61.8 to 256.5
C_6H_5Cl	Chlorobenzene	10098.0	8.50000	−35.0 to −15.0
		9067.3	7.71753	−13.0 to 349.8
C_6H_5ClO	3-Chlorophenol	11979.7	8.27628	44.2 to 214.0
$C_6H_5ClO_2S$	Benzenesulfonylchloride	12621.0	8.15333	65.9 to 251.5
$C_6H_5Cl_3Si$	Trichlorophenylsilane	11385.9	8.15335	33.0 to 201.0
$C_6H_5F_3Si$	Trifluorophenylsilane	9171.6	8.30502	−31.0 to 98.3
$C_6H_5NO_2$	Nitrobenzene	12168.2	8.41626	44.4 to 210.6
C_6H_6	1,5-Hexadiene-3-yne	8288.0	7.97728	−45.1 to 84.0
C_6H_6ClN	2-Chloroaniline	12441.0	8.56945	46.3 to 208.8
C_6H_6ClN	4-Chloroaniline	12832.8	8.46103	59.3 to 230.5
$C_6H_6N_2O_2$	2-Nitroaniline	15284.0	8.86838	104.0 to 284.5
$C_6H_6N_2O_2$	4-Nitroaniline	17220.2	9.04187	142.4 to 336.0
$C_6H_6O_2$	Pyrocatechol	13779.7	8.69431	104.0 to 245.5
$C_6H_6O_2$	Hydroquinone	18734.0	10.30930	132.4 to 286.2
C_6H_6S	Benzenethiol	11320.1	8.50439	18.6 to 168.0
C_6H_7N	2-Picoline	9933.2	8.29091	−11.1 to 128.8
C_6H_8	1,3-Cyclohexadiene	—	—	—
$C_6H_8N_2$	1,3-Phenylenediamine	14761.1	8.67082	99.8 to 285.5
$C_6H_8O_3$	α-Methylglutaric anhydride	14204.9	8.48075	93.8 to 282.5
$C_6H_8O_4$	Dimethyl maleate	12615.7	8.65944	45.7 to 205.0
C_6H_{10}	1,5-Hexadiene	—	—	—
$C_6H_{10}Cl_2Si$	Diallyldichlorosilane	10462.8	8.12150	9.5 to 165.3
$C_6H_{10}O$	Mesityl oxide	10109.4	8.38370	−8.7 to 130.0
$C_6H_{10}O_3$	Propionic anhydride	11572.6	8.63689	20.6 to 167.0
$C_6H_{10}0_3$	Methyl levulinate	12249.8	8.58030	39.8 to 197.7
$C_6H_{10}O_4$	Diethyl oxalate	14016.9	9.57295	47.4 to 185.7
$C_6H_{10}O_5$	Dimethyl-1-malate	14127.6	8.87979	75.4 to 242.6
$C_6H_{10}O_6$	Dimethyl-d1-tartrate	14999.1	8.78815	100.4 to 282.0
$C_6H_{11}BrO_2$	Ethyl α-bromoisobutyrate	10635.8	8.22167	10.6 to 163.6
$C_6H_{11}N$	Capronitrile	10492.3	8.14094	9.2 to 163.7
C_6H_{12}	2-Hexene	—	—	—
C_6H_{12}	Methylcyclopentane	7940.0	7.94547	−53.7 to 71.8

Vapor Pressures of Organic Compounds (continued)

FORMULA	NAME	A	B	RANGE
$C_6H_{12}Cl_2O_2$	bis(2-Chloroethyl)acetal	13497.1	8.95616	56.2 to 212.6
$C_6H_{12}O$	4-Methyl-2-pentanone	11669.6	9.40765	−1.4 to 119.0
$C_6H_{12}O$	Allyl isopropyl ether	8637.5	8.24287	−43.7 to 79.5
$C_6H_{12}O_2$	Caproic Acid	16189.4	10.43146	71.4 to 202.0
$C_6H_{12}O_2$	4-Hydroxy-4-methyl-2-pentan-one	11718.8	8.70338	22.0 to 167.9
$C_6H_{12}O_2$	Methyl isovalerate	9567.5	8.26669	−19.2 to 116.7
$C_6H_{12}O_2$	Ethyl isobutyrate	8945.7	7.94435	−24.3 to 280.0
$C_6H_{12}O_2$	n-Butyl acetate	—	—	—
$C_6H_{12}O_2$	n-Amyl formate	—	—	—
$C_6H_{12}O_3$	Paraformaldehyde	10348.2	8.59821	−9.4 to 124.0
C_6H_{14}	2-Methylpentane	7676.6	7.94463	−60.9 to 60.3
C_6H_{14}	2,2-Dimethylbutane	7271.0	7.84134	−69.3 to 49.7
$C_6H_{14}O$	1-Hexanol	12708.5	9.36761	24.4 to 157.0
$C_6H_{14}O$	3-Hexanol	11157.9	8.86051	2.5 to 135.5
$C_6H_{14}O$	2-Methyl-2-pentanol	11132.0	9.08500	−4.5 to 121.1
$C_6H_{14}O$	Ethyl butyl ether	—	—	—
$C_6H_{14}O$	Diisopropyl ether	777.3	7.90466	−57.0 to 67.5
$C_6H_{14}O_2$	1,2-Diethoxyethane	8102.6	7.43017	−33.5 to 119.5
$C_6H_{14}O_3$	Diethyleneglycol diethyl ether	12669.9	8.70561	45.3 to 201.9
$C_6H_{14}O_4$	Triethyleneglycol	17097.1	9.65286	114.0 to 278.3
$C_6H_{15}N$	Di-n-Propylamine	—	—	—
$C_6H_{15}B$	Triethylboron	2535.0	5.48687	−148.0 to −56.2
$C_6H_{15}O_4P$	Triethyl phosphate	11549.9	8.09039	39.6 to 211.0
$C_6H_{16}O_2Si$	Diethoxydimethylsilane	9758.2	8.43234	−19.1 to 113.5
$C_6H_{16}Sn$	Trimethylpropyltin	9659.6	8.13759	−12.0 to 131.7
$C_6H_{18}O_3Si_3$	Hexamethylcyclotrisiloxane	10503.3	8.51822	67.5 to 134.0
$C_7H_4ClF_3$	2-Chloro-α,α,α-trifluoroto-luene	10016.9	8.04377	.0 to 152.2
C_7H_5BrO	Benzoyl bromide	12070.8	8.25941	47.0 to 218.5
$C_7H_5Cl_3$	α,α,α-Trichlorotoluene	12168.6	8.36107	45.8 to 213.5
C_7H_5N	Benzonitrile	11341.0	8.23976	28.2 to 190.6
C_7H_5NO	Phenyl isocyanate	10556.7	8.15905	10.6 to 165.6
$C_7H_5NO_3$	3-Nitrobenzaldehyde	14726.9	8.73093	96.2 to 278.3
$C_7H_6Cl_2$	α,α-Dichlorotoluene	11075.9	7.87169	35.4 to 214.0
$C_7H_6O_2$	Benzoic acid	15253.3	9.03300	60.0 to 110.0
		16295.1	9.74136	96.0 to 249.2
$C_7H_6O_2$	4-Hydroxybenzaldehyde	16043.4	8.91213	121.2 to 310.0
C_7H_7Br	α-Bromotoluene	11360.4	8.15459	32.2 to 198.5
C_7H_7Br	3-Bromotoluene	10537.1	7.88656	14.8 to 183.7
C_7H_7BrO	4-Bromoanisole	12075.4	8.21777	48.8 to 223.0
C_7H_7Cl	2-Chlorotoluene	10279.3	8.09176	5.4 to 159.3
C_7H_7Cl	4-Chlorotoluene	10151.7	7.98836	5.5 to 162.3
C_7H_7F	3-Fluorotoluene	9251.8	8.10104	−22.4 to 116.0
C_7H_7I	2-Iodotoluene	11380.7	8.03523	37.2 to 211.0
$C_7H_7NO_2$	3-Nitrotoluene	11831.1	7.99921	50.2 to 231.9

(continued)

Vapor Pressures of Organic Compounds (continued)

FORMULA	NAME	A	B	RANGE
C_7H_8	Toluene	9368.5	8.33000	−92.0 to −15.0
		8580.5	7.71939	−26.7 to 319.0
$C_7H_8Cl_2Si$	Dichloromethylphenylislane	11464.7	8.13811	35.7 to 205.5
C_7H_8O	Anisole	10440.9	8.2214	5.4 to 155.5
C_7H_8O	o-Cresol	12487.3	8.79085	38.2 to 190.8
C_7H_8O	p-Cresol	13611.7	9.19055	53.0 to 201.8
$C_7H_8O_2$	2-Methoxyphenol	13425.8	9.02729	52.4 to 205.0
C_7H_9N	2,6-Dimethylpyridine	—	—	—
C_7H_9N	N-Methylaniline	11982.3	8.49145	36.0 to 195.5
C_7H_9N	3-Toluidine	12104.1	8.44037	41.0 to 203.3
C_7H_9NO	2-Methoxyaniline	13684.6	8.97119	61.0 to 218.5
$C_7H_{10}N_2$	4-Tolylhydrazine	15063.1	9.28676	82.0 to 242.0
$C_7H_{10}O_4$	Dimethyl citraconate	12917.3	8.72822	50.8 to 210.5
$C_7H_{10}O_4$	trans-Dimethyl mesaconate	12688.1	8.68097	46.8 to 206.0
$C_7H_{12}O_2$	Butyl acrylate	10194.0	8.20480	−0.5 to 147.4
$C_7H_{12}O_4$	Pimelic acid	19840.8	9.93471	163.4 to 342.1
$C_7H_{13}ClO$	Enanthyl chloride	15242.7	10.85555	34.2 to 145.0
C_7H_{14}	Ethylcyclopentane	8797.7	8.02549	−32.2 to 103.4
C_7H_{14}	1-Heptene	—	—	—
C_7H_{14}	Methylcyclohexane	8549.2	7.90976	−35.9 to 100.9
$C_7H_{14}O$	2-Heptanone	12478.9	9.30564	19.3 to 150.2
$C_7H_{14}O$	2,5-Dimethyl-3-pentanone	12266.9	9.64525	5.2 to 123.7
$C_7H_{14}O_2$	Methyl caproate	10676.8	8.39487	5.0 to 150.0
$C_7H_{14}O_2$	Propyl butyrate	10283.7	8.30515	−1.6 to 142.7
$C_7H_{14}O_2$	Isopropyl isobutyrate	9717.6	8.30764	−16.3 to 120.5
$C_7H_{14}O_2$	Isoamyl acetate	10494.9	8.42571	0.0 to 142.0
C_7H_{16}	n-Heptane	8928.8	8.25850	−63.0 to −40.0
		8409.6	7.78658	−34.0 to 247.5
C_7H_{16}	3-Methylhexane	8596.3	8.06547	−39.0 to 91.9
C_7H_{16}	2,2-Dimethylpentane	8106.7	7.94656	−49.0 to 79.2
C_7H_{16}	2,4-Dimethylpentane	8167.1	7.96137	−48.0 to 80.5
C_7H_{16}	2,2,3-Trimethylbutane	7767.1	7.69091	−18.8 to 80.9
$C_7H_{16}O_3$	Triethyl orthoformate	10935.0	8.60358	5.5 to 146.0
$C_7H_{18}Si$	Butyltrimethylsilane	9206.0	8.09332	−23.4 to 115.0
$C_8H_4Cl_2O_2$	Phthaloyl chloride	13716.0	8.35436	86.3 to 275.8
$C_8H_5Cl_2N$	α,α-Dichlorophenylacetonitrile	12829.9	8.54132	56.0 to 223.5
$C_8H_6Cl_2$	2,3-Dichlorostyrene	12827.2	8.41052	61.0 to 235.0
$C_8H_6Cl_2$	2,5-Dichlorostyrene	12592.5	8.39888	55.5 to 227.0
$C_8H_6Cl_2$	3,4-Dichlorostyrene	12626.5	8.37580	57.2 to 230.0
$C_8H_6Cl_4$	3,4,5,6-Tetrachloro-1,2-xylene	14763.1	8.79947	94.4 to 273.5
$C_8H_6O_2$	Phenylglyoxal	13731.6	9.32110	75.0 to 193.5
$C_8H_6O_3$	Piperonal	14425.5	8.77164	87.0 to 263.0
C_8H_7ClO	Phenylacetyl chloride	12626.1	8.61338	48.0 to 210.0
C_8H_7N	4-Tolunitrile	11562.8	8.04893	42.5 to 217.6
C_8H_7N	2-Tolyl isocyanide	11303.3	8.31275	25.2 to 183.5
C_8H_7NS	2-Methylbenzothiazole	14492.3	9.23950	70.0 to 225.5
$C_8H_8Br_2$	(1,2-Dibromoethyl)-benzene	14874.7	9.06312	86.0 to 254.0

Vapor Pressures of Organic Compounds (continued)

FORMULA	NAME	A	B	RANGE
$C_8H_8Cl_2$	1,2-Dichloro-4-ethyl-benzene	11711.5	7.99481	47.0 to 226.6
C_8H_8O	Acetophenone	11731.5	8.29324	371. to 202.4
$C_8H_8O_2$	Phenylacetic acid	15568.7	9.20612	97.0 to 265.5
$C_8H_8O_2$	Methyl benzoate	12077.2	8.50991	39.0 to 199.5
$C_8H_8O_3$	Vanillin	15703.2	9.03139	107.0 to 285.0
C_8H_9Br	2-Bromo-1,4-xylene	11603.7	8.19286	37.5 to 206.7
C_8H_9Br	(2-Bromoethyl)benzene	12152.5	8.29376	48.0 to 219.0
C_8H_9Cl	1-Chloro-3-ethylbenzene	10724.1	8.05722	18.6 to 181.1
C_8H_9ClO	1-Chloro-2-ethoxybenzene	12411.1	8.52952	45.8 to 208.0
$C_8H_{10}Cl_2Si$	Dichlorophenylethylsilane	11895.1	8.10018	48.5 to 225.5
$C_8H_9NO_2$	Methyl anthranilate	13186.3	8.23967	77.6 to 266.5
C_8H_{10}	Ethylbenzene	9301.3	7.80947	−9.8 to 326.5
C_8H_{10}	m-Xylene	9904.2	8.16704	−6.9 to 139.1
$C_8H_{10}Cl_2OSi$	Dichloroethoxyphenylsilane	12516.5	8.42845	52.4 to 222.2
$C_8H_{10}O$	2-Ethylphenol	12516.7	8.58694	46.2 to 207.5
$C_8H_{10}O$	4-Ethylphenol	13437.9	8.85499	59.3 to 219.0
$C_8H_{10}O$	2,3-Xylenol	13106.9	8.71884	56.0 to 218.0
$C_8H_{10}O$	2,5-Xylenol	13130.2	8.86726	51.8 to 211.5
$C_8H_{10}O$	3,5-Xylenol	13767.7	8.99775	62.0 to 219.5
$C_8H_{10}O$	α-Methyl benzyl alcohol	13087.4	8.90645	49.0 to 204.0
$C_8H_{10}O_2$	4,6-Dimethylresorcinol	12433.1	8.45696	49.0 to 215.0
$C_8H_{10}O_6$	Diethyl dioxosuccinate	13973.3	8.92007	70.0 to 233.5
$C_8H_{11}N$	N-Ethylaniline	11817.0	8.29268	38.5 to 204.0
$C_8H_{11}N$	4-Ethylaniline	12679.9	8.54135	52.0 to 217.4
$C_8H_{11}N$	2,6-Xylidine	11742.6	8.11428	44.0 to 217.9
$C_8H_{11}NO$	2-Anilinoethanol	15643.2	9.08123	104.0 to 279.6
$C_8H_{12}Cl_2O_5$	Diethyleneglycol-bischloroace-tate	19830.5	10.25225	148.3 to 313.0
$C_8H_{12}O_4$	Diethyl fumarate	12747.4	8.55490	53.2 to 218.5
$C_8H_{14}O_3$	Ethyl-α-ethylacetoacetate	12344.2	8.62223	40.5 to 198.0
$C_8H_{14}O_4$	Isopropyl levulinate	12689.6	8.66996	48.0 to 208.2
$C_8H_{14}O_4$	Diisopropyl oxalate	12949.3	8.96498	43.2 to 193.5
$C_8H_{14}O_4$	Diethyl isosuccinate	12087.6	8.46649	39.8 to 201.3
$C_8H_{14}O_5$	Diethyl malate	14202.9	8.78569	80.7 to 253.4
$C_8H_{14}O_6$	Diethyl-d-tartrate	15517.8	8.86286	192.0 to 280.0
$C_8H_{15}N$	n-Caprylonitrile	12221.8	8.52597	43.0 to 204.5
C_8H_{16}	1-Octene	—	—	—
C_8H_{16}	2-Methyl-2-heptene	9643.8	8.22861	−16.1 to 122.5
C_8H_{16}	cis-1,2-Dimethylcyclohexane	9364.9	8.00115	−15.9 to 129.7
C_8H_{16}	cis-1,3-Dimethylcyclohexane	9232.6	7.99243	−19.4 to 124.4
C_8H_{16}	cis-1,4-Dimethylcyclohexane	9188.9	7.96855	−20.0 to 124.3
C_8H_{16}	Ethylcyclohexane	9441.2	8.01137	−14.5 to 131.8
$C_8H_{16}O$	Cyclohexaneethanol	13152.4	8.90406	50.4 to 205.4
$C_8H_{16}O$	6-Methyl-5-hepten-2-ol	13999.1	9.71151	41.9 to 174.3
$C_8H_{16}O$	2,2,4-Trimethyl-3-pentanone	12854.6	9.781180	14.7 to 135.0
$C_8H_{16}O_2$	Ethyl isocaproate	10826.7	8.35435	11.0 to 160.4

(continued)

Vapor Pressures of Organic Compounds (continued)

FORMULA	NAME	A	B	RANGE
$C_8H_{16}O_2$	Isobutyl butyrate	10283.9	8.11142	4.6 to 156.9
$C_8H_{16}O_2$	Amylisopropionate	10567.2	8.21854	8.5 to 160.2
$C_8H_{17}I$	1-Iodooctane	11625.1	7.98981	45.8 to 225.5
C_8H_{18}	Octane	9221.0	7.89401	−14.0 to 291.4
C_8H_{18}	3-Methylheptane	9432.0	8.17940	−19.8 to 118.9
C_8H_{18}	2,2-Dimethylhexane	8927.8	8.05253	−29.7 to 106.8
C_8H_{18}	2,4-Dimethylhexane	9086.6	8.11387	−26.9 to 109.4
C_8H_{18}	3,3-Dimethylhexane	9065.2	8.06894	−25.8 to 112.0
C_8H_{18}	3-Ethylhexane	9416.3	8.17726	−20.0 to 118.5
C_8H_{18}	2,2,4-Trimethylpentane	8548.0	7.93485	−36.5 to 99.2
C_8H_{18}	2,3,4-Trimethylpentane	8988.2	7.99709	−26.3 to 113.5
C_8H_{18}	3-Methyl-3-ethylpentane	9028.7	7.95741	−23.9 to 118.3
$C_8H_{18}N_2$	Tetramethylpiperazine	11187.5	8.29203	23.7 to 183.5
C_8H_{18}	2-Octanol	12468.4	8.92992	32.8 to 178.5
$C_8H_{18}O$	Methyl heptyl ether	—	—	—
$C_8H_{18}O_3$	Diethylene glycol butyl ether	14127.0	9.08847	70.0 to 231.2
$C_8H_{18}S$	Di-n-butyl sulfide	11183.6	8.23456	21.7 to 182.0
$C_8H_{19}N$	Diisobutylamine	10058.3	8.22793	−5.1 to 139.5
$C_8H_{20}Pb$	Tetraethyllead	12959.7	9.12067	38.4 to 183.0
$C_8H_{20}Si$	Tetraethylsilane	9893.0	7.97412	−1.0 to 153.0
$C_8H_{22}O_3Si_2$	1,3-Diethoxytetramethyldisiloxane	11261.9	8.57664	14.8 to 160.7
$C_8H_{24}O_2Si_3$	Octamethyltrisiloxane	10956.0	8.56049	7.4 to 150.2
$C_9H_6O_2$	Coumarin	15202.7	8.78096	106.0 to 291.0
C_9H_7N	Isoquinoline	12847.6	8.36373	63.5 to 240.5
C_9H_8O	Cinnamyladehyde	14048.4	8.80198	76.1 to 246.0
C_9H_9N	Skatole	15232.7	9.06571	95.0 to 266.2
C_9H_{10}	α-Methyl styrene	10214.6	7.95975	7.4 to 165.4
C_9H_{10}	2-Methyl styrene	—	—	—
C_9H_{10}	4-Methyl styrene	10724.2	8.13090	16.0 to 175.0
$C_9H_{10}O$	Cinnamyl alcohol	13421.6	8.50124	72.6 to 250.0
$C_9H_{10}O$	2-Vinylanisole	12756.4	8.86645	41.9 to 194.0
$C_9H_{10}O$	4-Vinylanisole	12554.7	8.64027	45.2 to 204.5
$C_9H_{10}O_2$	Ethyl benzoate	11981.5	8.27959	44.0 to 213.4
$C_9H_{10}O_2$	Ethyl salicylate	13030.1	8.53813	61.2 to 231.5
$C_9H_{11}NO_2$	Ethyl carbanilate	19791.8	11.37309	107.8 to 237.0
C_9H_{12}	1,2,4-Trimethylbenzene	10710.2	8.20901	13.6 to 169.2
C_9H_{12}	o-Ethyl toluene	10488.8	8.14103	9.4 to 165.1
C_9H_{12}	p-Ethyl toluene	10461.1	8.17526	7.6 to 162.0
C_9H_{12}	N-Propylbenzene	10424.1	8.18588	6.3 to 159.2
$C_9H_{12}O$	3-Ethylanisole	11616.7	8.30303	33.7 to 196.5
$C_9H_{12}O$	3-Phenyl-1-propanol	14493.9	9.12735	74.7 to 235.0
$C_9H_{12}O$	3-Isopropylphenol	13292.2	8.68728	62.0 to 228.0
$C_9H_{12}O$	Benzyl ethyl ether	11315.5	8.29683	26.0 to 185.0
$C_9H_{13}N$	2,4,5-Trimethylaniline	13975.0	8.98182	68.4 to 234.5
$C_9H_{13}N$	N,N-Dimethyl-4-toluidine	12738.4	8.70583	50.1 to 209.5
$C_9H_{14}O$	Phorone	12557.2	8.72908	42.0 to 197.2

Vapor Pressures of Organic Compounds (continued)

FORMULA	NAME	A	B	RANGE
$C_9H_{14}O_4$	cis-Diethyl citraconate	12913.2	8.49764	59.8 to 230.3
$C_9H_{14}O_4$	Diethyl mesaconate	13326.1	8.68826	62.8 to 229.0
$C_9H_{16}O_3$	Isobutyl levulinate	13571.2	8.81607	65.0 to 229.9
$C_9H_{16}O_4$	Diethyl ethylmalonate	12842.0	8.68722	50.8 to 211.5
$C_9H_{18}O$	2-Nonanone	11529.5	8.27996	32.1 to 195.0
$C_9H_{18}O$	Azelaldehyde	12143.4	8.69138	33.0 to 185.0
$C_9H_{18}O_2$	Methyl caprylate	11914.9	8.47845	34.2 to 193.0
$C_9H_{18}O_2$	Isoamyl butyrate	11104.5	8.27274	21.2 to 178.6
$C_9H_{19}I$	1-Iodononane	14853.0	9.48218	70.0 to 219.5
C_9H_{20}	2,6-Dimethylheptane	—	—	—
C_9H_{20}	3-Methyloctane	—	—	—
$C_9H_{20}O$	Diisobutyl carbinol	—	—	—
$C_9H_{20}O_4$	Tripropyleneglycol	15291.4	9.07197	96.0 to 267.2
$C_9H_{22}Si$	Triethylpropylsilane	10709.3	8.14477	15.2 to 173.0
$C_{10}H_7Cl$	1-Chloroanphthalene	13570.5	8.51464	80.6 to 259.3
$C_{10}H_8$	Naphthalene	17065.2	11.45000	.0 to 80.0
		12311.6	8.41308	52.6 to 217.9
$C_{10}H_8O$	1-Naphthol	14205.6	8.47666	94.0 to 282.5
$C_{10}H_9N$	1-Naphthylamine	14529.5	8.42790	104.3 to 300.8
$C_{10}H_9N$	2-Methylquinoline	14154.0	8.89035	75.3 to 246.5
$C_{10}H_{10}O$	4-Phenyl-3-buten-2-one	13913.9	8.58607	81.7 to 261.0
$C_{10}H_{10}O_2$	Methyl cinnamate	13325.5	8.33608	77.4 to 263.0
$C_{10}H_{10}O_4$	1,2-Phenylene diacetate	14986.0	8.82415	98.0 to 278.0
$C_{10}H_{12}$	2,4-Dimethylstyrene	11454.0	8.16537	34.2 to 202.0
$C_{10}H_{12}$	3-Ethylstyrene	11285.7	8.21090	28.3 to 191.5
$C_{10}H_{12}$	Tetralin	11613.0	8.19495	38.0 to 207.2
$C_{10}H_{12}O$	4-Methylpropiophenone	12505.0	8.23624	59.6 to 238.5
$C_{10}H_{12}O$	Cuminal	12668.0	8.37450	58.0 to 232.0
$C_{10}H_{12}O_2$	Eugenol	13907.8	8.66713	78.4 to 253.5
$C_{10}H_{12}O_2$	Chavibetol	14527.7	8.91795	83.6 to 254.0
$C_{10}H_{12}O_3$	2-Phenoxyethyl acetate	14070.3	8.65284	82.6 to 259.7
$C_{10}H_{13}Cl_2O_2P$	4-tert-Butylphenyl dichloro-phosphate	13711.0	8.13871	96.0 to 299.0
$C_{10}H_{14}$	1,2,3,5-Tetramethylbenzene	12358.4	8.68024	40.6 to 197.9
$C_{10}H_{14}$	4-Ethyl-1,3-xylene	11070.4	8.18691	23.2 to 184.5
$C_{10}H_{14}$	2-Ethyl-1,4-xylene	11144.6	8.21724	24.1 to 185.0
$C_{10}H_{14}$	1,3-Diethylbenzene	10993.9	8.17745	21.7 to 182.2
$C_{10}H_{14}$	1-Methyl-2-isopropylbenzene	—	—	—
$C_{10}H_{14}$	N-Butylbenzene	11052.1	8.19417	22.7 to 183.1
$C_{10}H_{14}$	sec-Butylbenzene	11069.3	8.31801	18.6 to 173.5
$C_{10}H_{14}O$	Carvacrol	13765.7	8.78569	70.0 to 237.0
$C_{10}H_{14}O$	Cuminyl alcohol	13799.2	8.70169	74.2 to 246.6
$C_{10}H_{14}O$	2-Isopropyl-5-methylphenol	13352.8	8.66413	64.3 to 231.8
$C_{10}H_{14}O$	4-sec-Butylphenol	13690.2	8.70313	71.4 to 242.1
$C_{10}H_{14}O$	2-tert-Butylphenol	13112.3	8.71014	56.6 to 219.5
$C_{10}H_{14}N_2$	Nicotine	12337.1	8.07765	61.8 to 247.3

(continued)

Vapor Pressures of Organic Compounds (continued)

FORMULA	NAME	A	B	RANGE
$C_{10}H_{15}NO_2$	N-Phenyliminodiethanol	17482.1	9.14449	145.0 to 337.8
$C_{10}H_{16}$	Dipentene	10538.3	8.03646	14.0 to 174.6
$C_{10}H_{16}$	Myrcene	10704.8	8.16124	14.5 to 171.5
$C_{10}H_{16}$	α-Pinene	9813.6	7.89820	−1.0 to 155.0
$C_{10}H_{16}$	Terpenoline	12030.8	8.63342	32.3 to 185.0
$C_{10}H_{16}O$	d-Camphor	12800.9	8.79900	.0 to 180.0
		11978.0	8.35230	41.5 to 209.2
$C_{10}H_{16}O$	α-Citral	13255.5	8.67195	61.7 to 228.0
$C_{10}H_{16}O$	Pulegone	13395.4	8.95594	58.3 to 221.0
$C_{10}H_{16}OSi$	Ethoxydimethylphenylsilane	11718.6	8.30223	36.3 to 199.5
$C_{10}H_{16}O_2$	Diosphenol	13644.0	8.79104	66.7 to 232.0
$C_{10}H_{18}$	cis-Decalin	10515.4	7.79754	22.5 to 194.6
$C_{10}H_{18}O$	d-Citronellal	12305.1	8.50036	44.0 to 206.5
$C_{10}H_{18}O$	Dihydrocarveol	13698.5	8.90339	63.9 to 225.0
$C_{10}H_{18}O$	Geraniol	14060.7	8.99977	69.2 to 230.0
$C_{10}H_{18}O$	Nerol	13366.1	8.73601	61.7 to 226.0
$C_{10}H_{18}O_2$	Citronellic acid	16455.4	9.67047	99.5 to 257.0
$C_{10}H_{18}O_3$	Isoamyl levulinate	13867.9	8.72224	75.6 to 247.9
$C_{10}H_{18}O_4$	Diethyl adipate	14240.6	8.89644	74.0 to 240.0
$C_{10}H_{18}O_4$	Dipropyl succinate	13975.7	8.72228	77.5 to 250.8
$C_{10}H_{18}O_6$	Dipropyl-d-tartrate	15754.0	8.87556	115.6 to 303.0
$C_{19}H_{19}N$	Camphylamine	13224.1	9.07553	45.3 to 195.0
$C_{10}H_{20}$	1-Decene	10233.3	7.84140	14.7 to 192.0
$C_{10}H_{20}O$	Citronellol	14214.1	9.163045	66.4 to 221.5
$C_{10}H_{20}O$	1-Menthol	13475.3	8.96520	56.0 to 212.0
$C_{10}H_{20}O_2$	Capric acid	19372.6	10.87266	125.0 to 268.4
$C_{10}H_{20}$	n-Decane	10912.0	8.24808	17.1 to 173.0
$C_{10}H_{22}$	2,7-Dimethyloctane	10339.3	8.13866	6.3 to 159.7
$C_{10}H_{22}O_2$	2-Butyl-2-ethylbutane-1,3-diol	15833.7	9.43777	94.1 to 255.0
$C_{10}H_{22}O_3$	Dipropylene glycol monobutyl ether	13721.1	8.90600	64.7 to 227.0
$C_{10}H_{24}Si$	Heptyltrimethylsilane	10987.3	8.15355	22.3 to 184.0
$C_{10}H_{28}O_4Si_3$	1,5-Diethoxyhexamethyltrisiloxane	12586.4	8.76191	41.8 to 196.6
$C_{10}H_{30}O_5Si_5$	Decamethylcyclopentasiloxane	12272.1	8.45329	45.2 to 210.0
$C_{11}H_8O_2$	2-Naphthoic acid	22630.8	11.39201	160.8 to 308.5
$C_{11}H_{12}O_2$	Ethyl-trans-cinnamate	13639.9	8.35185	87.6 to 271.0
$C_{11}H_{12}O_3$	Ethyl benzoylacetate	17115.4	9.82429	107.6 to 265.0
$C_{11}H_{14}$	1-Phenylpentane	—	—	—
$C_{11}H_{14}$	2,4,6-Trimethylstyrene	11588.8	8.17085	37.5 to 207.0
$C_{11}H_{14}O$	Isobutyrophenone	12878.8	8.50887	58.3 to 228.0
$C_{11}H_{14}O$	2,3,5-Trimethylacetophenone	14283.6	8.88554	79.0 to 247.5
$C_{11}H_{14}O_2$	4-Allylveratrole	15027.1	9.18800	85.0 to 248.0
$C_{11}H_{16}$	3,5-Diethyltoluene	11167.4	8.08310	31.8 to 199.0
$C_{11}H_{16}$	1,3,5-Trimethyl-2-ethylbenzene	11677.3	8.20078	38.8 to 208.0
$C_{11}H_{16}$	4-Ethylcumene	11425.6	8.22315	31.5 to 195.8
$C_{11}H_{16}O$	4-tert-Butyl-2-cresol	13798.1	8.69925	74.3 to 247.0

Vapor Pressures of Organic Compounds (continued)

FORMULA	NAME	A	B	RANGE
$C_{11}H_{16}O$	4-tert-Amylphenol	13154.3	8.21575	109.8 to 266.0
$C_{11}H_{18}O_2$	Bornyl formate	12276.0	8.39862	47.0 to 214.0
$C_{11}H_{18}O_2$	Neryl formate	12959.3	8.59120	57.3 to 224.5
$C_{11}H_{18}O_5$	Deithyl-gamma-oxoazelate	17543.6	9.73815	121.0 to 286.0
$C_{11}H_{20}O_2$	Menthyl formate	12077.7	8.25673	47.3 to 219.0
$C_{11}H_{20}O_2$	Octyl acrylate	12957.5	8.54078	58.5 to 227.0
$C_{11}H_{22}O$	Hendecan-2-one	14353.5	9.20552	68.2 to 224.0
$C_{11}H_{22}O_2$	Hendecanoic acid	14689.9	8.59430	101.4 to 290.0
$C_{11}H_{24}O$	Hendecan-2-ol	14216.2	9.04349	71.1 to 232.0
$C_{11}H_{26}Si$	Amyltriethylsilane	11859.7	8.25034	41.8 to 211.0
$C_{12}H_9BrO$	2-Bromo-4-Phenylphenol	13589.9	7.96634	100.0 to 311.0
$C_{12}H_9Cl$	4-Chlorobiphenyl	14017.4	8.30014	96.4 to 292.9
$C_{12}H_9ClO$	2-Chloro-6-phenylphenol	15508.4	8.63289	119.8 to 317.0
$C_{12}H_9N$	Carbazole	15421.6	8.25192	248.2 to 354.8
$C_{12}H_{10}$	Diphenyl	12910.0	8.21858	70.6 to 254.9
$C_{12}H_{10}Cl_2Si$	Dichlorodiphenylsilane	14968.5	8.57113	109.6 to 304.0
$C_{12}H_{10}N_2$	Azobenzene	14786.7	8.60058	103.5 to 293.0
$C_{12}H_{10}O$	2-Acetonaphthone	16496.7	9.16387	120.2 to 301.0
$C_{12}H_{10}O$	2-Phenylphenol	15397.8	9.01537	100.0 to 275.0
$C_{12}H_{10}S$	Diphenyl sulfide	13974.8	8.29321	96.1 to 292.5
$C_{12}H_{10}Se$	Diphenyl selenide	14603.4	8.44942	105.7 to 301.5
$C_{12}H_{12}$	1-Ethylnaphthalene	12751.3	8.13879	70.0 to 258.1
$C_{12}H_{14}N_2O_5$	2-Cyclohexyl-4-,6-dinitrophe-nol	19100.0	10.29306	132.8 to 291.5
$C_{12}H_{14}O_4$	Apiole	16881.7	9.49899	116.0 to 285.0
$C_{12}H_{16}$	2,5-Diethylstyrene	12150.3	8.24729	49.7 to 223.0
$C_{12}H_{16}O_2$	Isoamyl benzoate	12782.9	8.08037	72.0 to 262.0
$C_{12}H_{18}$	1,2,4-Triethylbenzene	11957.9	8.21406	46.0 to 218.0
$C_{12}H_{18}$	1,3,5-Triethylbenzene	—	—	—
$C_{12}H_{18}$	1,3-Diisopropylbenzene	11498.9	8.18607	34.7 to 202.0
$C_{12}H_{18}O$	4-tert-Butyl-2,5-xylenol	14477.9	8.75202	88.2 to 265.3
$C_{12}H_{18}O$	6-tert-Butyl-2,4-xylenol	13882.4	8.81691	70.3 to 236.5
$C_{12}H_{20}O_2$	d-Bornyl acetate	11838.7	8.10804	46.9 to 223.0
$C_{12}H_{20}O_2$	Linalyl acetate	12910.6	8.62394	55.4 to 220.0
$C_{12}H_{20}O_7$	Triethyl citrate	14818.4	8.59559	107.0 to 294.0
$C_{12}H_{22}O_2$	Citronellyl acetate	15781.3	9.92960	74.7 to 217.0
$C_{12}H_{22}O_4$	Dimethyl sebacate	14861.3	8.56294	104.0 to 293.5
$C_{12}H_{22}O_6$	Diisobutyl-d-tartrate	14874.9	8.33837	117.8 to 324.0
$C_{12}H_{24}$	Triisobutylene	10790.4	8.13865	18.0 to 179.0
$C_{12}H_{24}O$	Lauraldehyde	13644.2	8.51004	77.7 to 257.0
$C_{12}H_{26}$	n-Dodecane	11857.7	8.15099	47.7 to 345.8
$C_{12}H_{26}O_4$	Tripropylene glylcol monoiso-propyl ether	14171.5	8.72193	82.4 to 256.6
$C_{12}H_{27}N$	Dodecylamine	14836.4	9.11239	82.8 to 248.0
$C_{12}H_{34}O_5Si_4$	1,7-iethoxyoctamethylte-trasiloxane	14095.9	9.06241	67.7 to 227.5

(continued)

Vapor Pressures of Organic Compounds (continued)

FORMULA	NAME	A	B	RANGE
$C_{12}H_{36}O_6Si_6$	Dodecamethylcyclohexasiloxane	13760.6	8.84797	67.3 to 236.0
$C_{13}H_{10}$	Fluorene	13682.8	8.12889	129.3 to 295.0
$C_{13}H_{10}O_2$	Phenyl benzoate	14181.7	8.17562	106.8 to 314.0
$C_{13}H_{12}$	Diphenylmethane	13089.4	8.21592	76.0 to 264.5
$C_{13}H_{12}O$	Benzyl phenyl ether	14156.7	8.41281	95.4 to 287.0
$C_{13}H_{13}ClSi$	Chloromethyldiphenylsilane	14924.6	8.64717	105.0 to 295.0
$C_{13}H_{14}$	2-Isopropylnaphthalene	13036.9	8.17734	76.0 to 266.0
$C_{13}H_{20}$	Heptylbenzene	13535.4	8.73260	66.2 to 233.0
$C_{13}H_{22}O_2$	Bornyl propionate	13245.0	8.58676	64.6 to 235.0
$C_{13}H_{26}O_2$	Methyl laurate	14853.5	8.99167	87.8 to 190.8
$C_{13}H_{28}$	Tridecane	12991.3	8.49173	59.4 to 234.0
$C_{13}H_{30}Si$	Decyltrimethylsilane	13311.1	8.56522	67.4 to 240.0
$C_{14}H_8O_2$	Anthraquinone	21163.1	10.12792	190.0 to 379.9
$C_{14}H_{10}$	Anthracene	16823.6	8.70600	100.0 to 600.0
$C_{14}H_{10}O_2$	Benzil	15046.4	8.20173	128.4 to 347.0
$C_{14}H_{12}$	1,1-Diphenylethylene	13778.1	8.36908	87.4 to 277.0
$C_{14}H_{12}O$	Desoxybenzoin	15642.1	8.64353	123.3 to 321.0
$C_{14}H_{14}$	Dibenzyl	13387.6	8.14244	86.8 to 284.0
$C_{14}H_{15}O$	Dibenzylamine	16260.1	9.09432	118.3 to 300.0
$C_{14}H_{20}Cl_2$	1,2-Dichlorotetraethylbenzene	14629.0	8.46020	105.6 to 302.0
$C_{14}H_{20}O_3$	2-(4-tert-Butylphenoxy)ethyl acetate	16017.6	8.96789	118.0 to 304.4
$C_{14}H_{22}O$	2,4-Di-tert-butylphenol	14237.7	8.70909	84.5 to 260.8
$C_{14}H_{24}O_2$	Bornyl isobutyrate	13501.8	8.61288	70.0 to 243.0
$C_{14}H_{24}O_2$	Geranyl isobutyrate	15699.5	9.43949	90.7 to 251.0
$C_{14}H_{28}O$	2-Tetradecanone	15102.7	8.87971	99.3 to 278.0
$C_{14}H_{28}O_2$	Myristic acid	18380.1	9.66777	142.0 to 318.0
$C_{14}H_{30}$	Tetradecane	13750.0	8.62869	76.4 to 252.5
$C_{14}H_{32}Si$	Triethyloctylsilane	12954.8	8.18360	73.7 to 262.0
$C_{14}H_{42}O_5Si_6$	Tetradecamethylhexasiloxane	13800.0	8.71803	73.7 to 245.5
$C_{15}H_{14}O$	1,3-Diphenyl-2-propanone	15429.8	8.47844	125.5 to 330.5
$C_{15}H_{16}O_2$	4,4-Isopropylidenebisphenol	23254.0	10.90469	193.0 to 360.5
$C_{15}H_{18}OSi$	Ethoxymethyldiphenylsilane	16106.4	9.24867	109.0 to 282.0
$C_{15}H_{24}$	Cadiene	15518.3	9.08046	101.3 to 275.0
$C_{15}H_{24}O$	4,6-Di-tert-butyl-2-cresol	14006.9	8.53829	86.2 to 269.3
$C_{15}H_{26}O$	Champacol	14655.9	8.59916	100.0 to 288.0
$C_{15}H_{30}O_2$	Methyl myristate	16051.0	9.06656	115.0 to 295.8
$C_{15}H_{32}O_5$	Tetrapropylene glycol monoisopropyl ether	16494.6	9.25590	116.6 to 292.7
$C_{16}H_{14}O_2$	Benzyl cinnamate	20840.6	10.19782	173.8 to 350.0
$C_{16}H_{20}O_2Si$	Diethoxydiphenylsilane	15828.8	9.01457	111.5 to 296.0
$C_{16}H_{25}Cl$	Pentaethylchlorobenzene	13-07.3	8.24102	20.0 to 285.0
$C_{16}H_{26}O$	2,6-Di-tert-butyl-4-ethylphenol	14438.0	8.69446	89.1 to 286.6
$C_{16}H_{30}O$	Muscone	14722.5	8.24764	118.0 to 328.0
$C_{16}H_{32}$	1-Hexadecene	15634.7	9.13793	101.6 to 274.0
$C_{16}H_{32}O$	2-Hexadecanone	15194.4	8.58132	109.8 to 307.0

Vapor Pressures of Organic Compounds (continued)

FORMULA	NAME	A	B	RANGE
$C_{16}H_{32}O_2$	Palmitic acid	17603.6	9.03269	153.6 to 353.8
$C_{16}H_{34}O$	Cetyl alcohol	14483.4	8.01607	122.7 to 344.0
$C_{16}H_{36}Si$	Decyltriethylsilane	15393.7	8.83467	108.5 to 293.0
$C_{16}H_{48}O_6Si_7$	Hexadecamethylheptasiloxane	14841.5	8.87416	93.2 to 270.0
$C_{17}H_{10}O$	Benzanthrone	18309.6	8.01852	225.0 to 426.5
$C_{17}H_{24}O_2$	Menthyl benzoate	16804.5	9.28284	123.2 to 301.0
$C_{17}H_{34}O_2$	Methyl palmitate	17003.5	9.12564	134.3 to 202.0
$C_{17}H_{38}Si$	Tetradecyltrimethylsilane	16439.7	9.16224	120.0 to 300.0
$C_{18}H_{15}O_4P$	Triphenyl phosphate	19272.3	9.05278	193.5 to 413.5
$C_{18}H_{30}O$	2,4,6-Tri-tert-butylphenol	14703.7	8.73790	95.2 to 276.3
$C_{18}H_{34}O_2$	Elaidic acid	19538.0	9.59679	171.3 to 362.0
$C_{18}H_{36}O_2$	Stearic acid	19306.6	9.45747	173.7 to 370.0
$C_{18}H_{38}$	2-Methylheptadecane	16095.9	8.96073	119.8 to 306.5
$C_{18}H_{39}N$	Ethylcetylamine	15718.3	8.47133	133.2 to 342.0
$C_{18}H_{54}O_7Si_8$	Octadecamethyloctasiloxane	15270.3	8.83030	105.8 to 290.0
$C_{19}H_{40}$	Nonadecane	16497.3	8.89655	133.2 to 330.0
$C_{20}H_{43}N$	Diethylhexadecyclamine	15871.3	8.41918	139.8 to 355.0
$C_{20}H_{60}O_8Si_9$	Eicosamethylnonasiloxane	19522.9	10.23479	144.0 to 307.5
$C_{21}H_{44}$	Heneicosane	17702.2	9.08816	152.6 to 350.5
$C_{22}H_{42}O_2$	Brassidic acid	24085.7	10.92715	209.6 to 382.5
$C_{22}H_{56}O_9Si_{10}$	Docosamethyldecasiloxane	21878.6	11.03737	160.3 to 314.0
$C_{24}H_{50}$	Tetracosane	19642.5	9.40816	183.8 to 386.4
$C_{25}H_{52}$	Pentacosane	20815.9	9.73425	194.2 to 390.3
$C_{27}H_{33}O_4P$	Dicarvacryl-2-tolyl phosphate	24233.3	11.68778	180.2 to 330.0
$C_{28}H_{58}$	Octaosane	24144.2	10.57546	226.5 to 412.5
$C_{32}H_{34}ClO_4P$	Dicarvacryl-mono(6-chloro-2-xenyl)phosphate	25299.5	11.58067	204.2 to 361.0

Reprinted with permission from: Weast, *Handbook of Chemistry and Physics,* 53rd. Ed., CRC Press Inc., Boca Raton, FL., 1972.

Table A.2. Solubility Parameters of Various Liquids at 25°C

CODE	NAME	MOLAR VOLUME (V)	PARAMETERS HILDEBRANDS		
			δ_D	δ_P	δ_H
PARAFFIN HYDROCARBONS					
1	*n*-Butane	101.4	6.9	0	0
2	*n*-Pentane	116.2	7.1	0	0
3	*i*-Pentane	117.4	6.7	0	0
4	*n*-Hexane	131.6	7.3	0	0
5	*n*-Heptane	147.4	7.5	0	0
6	*n*-Octane	163.5	7.6	0	0
7	2,2,4-Trimethylpentane	166.1	7.0	0	0
8	*n*-Nonane	179.7	7.7	0	0
9	*n*-Decane	195.9	7.7	0	0
10	*n*-Dodecane	228.6	7.8	0	0
11	*n*-Hexadecane	294.1	8.0	0	0
12	*n*-Eicosane	359.8	8.1	0	0
13	Cyclohexane	108.7	8.2	0	0.1[a]
14	Methylcyclohexane	128.3	7.8	0	0.5
14.1	*cis*-Decahydronaphthalene	156.9	9.2	0	0
14.2	*trans*-Decahydronaphthalene	159.9	8.8	0	0
AROMATIC HYDROCARBONS					
15	Benzene	89.4	9.0	0[a]	1.0
16	Toluene	106.8	8.8	0.7	1.0
16.1	Naphthalene[b]	111.5	9.4	1.0	2.9
17	Styrene	115.6	9.1	0.5	2.0
18	*o*-Xylene	121.2	8.7	0.5	1.5
19	Ethylbenzene	123.1	8.7	0.3	0.7
19.1	1-Methylnaphthalene	138.8	10.1	0.4	2.3
20	Mesitylene	139.8	8.8	0	0.3
21	Tetrahydronaphthalene	136.0[a]	9.6[a]	1.0	1.4
21.1	Biphenyl	154.1	10.5	0.5	1.0
22	*p*-Diethylbenzene	156.9	8.8	0	0.3
HALOHYDRO-CARBONS					
23	Methyl chloride	55.4	7.5[a]	3.0	1.9
24	Methylene dichloride	63.9	8.9	3.1	3.0
24.1	Bromochloromethane	65.0	8.5	2.8	1.7
25	Chlorodifluoromethane	72.9	6.0	3.1	2.8
26	Dichlorofluoromethane	75.4	7.7	1.5	2.8
27	Ethyl bromide	76.9	8.1	3.9	2.5
27.1	1,1-Dichloroethylene	79.0	8.3	3.3	2.2
28	Ethylene dichloride	79.4	9.3[a]	3.6	2.0
28.1	Methylene di-iodide[c]	80.5	8.7	1.9	2.7
29	Chloroform	80.7	8.7	1.5	2.8
29.1	1,1-Dichloroethane	84.8	8.1	4.0	0.2

Table A.2. (Continued)

CODE	NAME	MOLAR VOLUME (V)	PARAMETERS HILDEBRANDS		
			δ_D	δ_P	δ_H
HALOHYDRO-CARBONS					
29.2	Ethylene dibromide	87.0	9.6	3.3	5.9
30	Bromoform	87.5	10.5[a]	2.0	3.0[a]
31	n-Propyl chloride	88.1	7.8	3.8	1.0
32	Trichloroethylene	90.2	8.8	1.5	2.6
33	Dichlorodifluoromethane	92.3	6.0	1.0	0
34	Trichlorofluoromethane	92.8	7.5	1.0	0
35	Bromotrifluoromethane	97.0	4.7	1.2	0
36	Carbon tetrachloride	97.1	8.7	0	0.3
37	1,1,1-Trichloroethane	100.4	8.3	2.1	1.0
38	Tetrachloroethylene	101.1	9.3	3.2[a]	1.4
39	Chlorobenzene	102.1	9.3	2.1	1.0
39.1	n-Butylchloride	104.9	8.0	2.7	1.0
39.2	1,1,2,2-Tetrachloroethane	105.2[a]	9.2	2.5	4.6
40	Bromobenzene	105.3	10.0	2.7	2.0
41	o-Dichlorobenzene	112.8	9.4	3.1	1.6
42	Benzyl chloride	115.0	9.2[a]	3.5	1.3
42.1	1,1,2,2-Tetrabromoethane[c]	116.8	11.1	2.5	4.0
43	1,2-Dichlorotetrafluoroethane[c]	117.0	6.2	0.9	0
44	1,1,2-Trichlorotrifluoroethane	119.2	7.2	0.8	0
45	Cycclohexyl chloride	121.3	8.5	2.7	1.0
46	1-Bromonaphthalene	140.0	9.9	1.5	2.0
47	Trichlorobiphenyl[d]	187.0	9.4	2.6	2.0
48	Perfluoromethylcyclohexane	196.0	6.1	0	0
49	Perfluorodimethylcyclohexane[d]	217.4	6.1	0	0
50	Perfluoro-n-heptane	227.3	5.9	0	0
ETHERS					
51	Furan	72.5	8.7	0.9	2.6
51.1	Epichlorhydrin	79.9	9.3	5.0	1.8
51.2	Tetrahydrofuran	81.7	8.2	2.8	3.9
51.3	1,4-dioxane	85.7	9.3	0.9	3.6
51.4	Methylal[c] $CH_2 (OCH)_2$	88.8	7.4	0.9	4.2
52	Diethyl ether	104.8	7.1	1.4	2.5
53	bis(2-Chloroethyl)ether	117.6	9.2	4.4	2.8[a]
53.1	Anisole[c]	119.1	8.7	2.0	3.3
53.2	di-(2-Methoxyethyl)ether	142.0	7.7	3.0	4.5
53.3	Dibenzyl ether[c]	192.7	8.5	1.8	3.6
53.4	di-(2-Chloro-i-propyl)ether[c]	146.0	9.3	4.0	2.5

(continued)

Table A.2. (Continued)

CODE	NAME	MOLAR VOLUME (V)	PARAMETERS HILDEBRANDS		
			δ_D	δ_P	δ_H
ETHERS					
54	bis-(m-Phenoxy-phenyl)ether	373.0	9.6	1.5	2.5
KETONES					
55	Acetone	74.0	7.6	5.1	3.4
56	Methyl ethyl ketone	90.1	7.8	4.4	2.5
57	Cyclohexanone	104.0	8.7	3.1	2.5
58	Diethyl ketone	106.4	7.7	3.7	2.3
58.1	Mesityl oxide	115.6	8.0	3.5	3.0
59	Acetophenone	117.4	9.6[a]	4.2	1.8
60	Methyl i-butyl ketone	125.8	7.5	3.0	2.0
61	Methyl i-amyl ketone	142.8	7.8	2.8	2.0
61.1	Isophorone	150.5	8.1	4.0	3.6
62	di-(i-Butyl)ketone	177.1	7.8	1.8	2.0
ALDEHYDES					
63	Acetaldehyde[c]	57.1	7.2	3.9	5.5
63.1	Furfuraldehyde	83.2	9.1[a]	7.3	2.5
64	Butyraldehyde	88.5	7.2	2.6[a]	3.4[a]
65	Benzaldehyde	101.5	9.5	3.6	2.6
ESTERS					
66	Ethylene carbonate	66.0	9.5	10.6	2.5
66.1	γ-Butyrolactone	76.8	9.3	8.1	3.6
66.2	Methyl acetate	79.7	7.6	3.5	3.7
67	Ethyl formate	80.2	7.6	4.1	4.1
67.1	Propylene carbonate	85.0	9.8	8.8	2.0
68	Ethyl chloroformate	95.6	7.6	4.9	3.3
69	Ethyl acetate	98.5	7.7[a]	2.6	3.5[a]
69.1	Trimethyl phosphate	99.9	8.2	7.8	5.0
70	Diethyl carbonate	121.0	8.1	1.5	3.0
71	Diethyl sulfate	131.5	7.7	7.2	3.5
72	n-Butyl acetate	132.5	7.7	1.8	3.1
72.1	i-Butyl acetate	133.5	7.4	1.8	3.1
72.2	2-Ethoxyethyl acetate	136.2	7.8	2.3	5.2
73	i-Amyl acetate	148.8	7.5	1.5	3.4
73.1	i-Butyl i-butyrate	163.0	7.4	1.4	2.9
74	Dimethyl phthalate	163.0	9.1[a]	5.3[a]	2.4
75	Ethyl cinnamate	166.8	9.0	4.0	2.0
75.1	Triethyl phosphate	171.0	8.2	5.6	4.5
76	Diethyl phthalate	198.0	8.6	4.7	2.2
76.1	di-n-Butyl phthalate	266.0	8.7[a]	4.2	2.0
76.2	n-Butyl benzyl phthalate	306.0	9.3	5.5	1.5
77	Tricresyl phosphate	316.0	9.3	6.0	2.2
78	tri-n-Butyl phosphate	345.0	8.0	3.1	2.1
79	i-Propyl palmitate[c]	330.0	7.0	1.9	1.8
79.1	di-n-Butyl sebacate	339.0	6.8	2.2	2.0

Table A.2. (Continued)

CODE	NAME	MOLAR VOLUME (V)	δ_D	δ_P	δ_H
ESTERS					
79.2	Methyl oleate[d]	340.0	7.1	1.9	1.8
79.3	Dioctyl phthalate	377.0	8.1	3.4	1.5
80	di-Butyl stearate[c]	382.0	71.0	1.8	1.7
NITROGEN COMPOUNDS					
81	Acetonitrile	52.6	7.5	8.8	3.0
81.1	Acrylonitrile	67.1	8.0	8.5	3.3
82	Propionitrile	70.9	7.5	7.0	2.7
83	Butyronitrile	87.0	7.5	6.1	2.5
84	Benzonitrile	102.6	8.5	4.4	1.6
85	Nitromethane	54.3	7.7	9.2	2.5
86	Nitroethane	71.5	7.8	7.6	2.2
87	2-Nitropropane	86.9	7.9	5.9	2.0
88	Nitrobenzene	102.7	9.8	4.2	2.0
89	Ethanolamine	60.2	8.4	7.6	10.4
89.1	Ethylene diamine	67.3	8.1	4.3	8.3
89.2	1,1-Dimethylhydrazine[c]	76.0	7.5	2.9	5.4
89.3	2-Pyrrolidone	76.4	9.5	8.5	5.5
90	Pyridine	80.9	9.3	4.3	2.9
91	n-Propylamine	83.0	8.3	2.4	4.2
92	Morpholine	87.1	9.2	2.4	4.5
93	Aniline	91.5	9.5	2.5	5.0
93.1	N-Methyl-2-pyrrolidone	96.5	8.8	6.0	3.5
94	n-Butylamine	99.0	7.9[a]	2.2[a]	3.9[a]
95	Diethylamine	103.2	7.3	1.1	3.0
95.1	Diethylenetriamine	108.0	8.2	6.5	7.0
96	Cyclohexylamine	115.2	8.5	1.5	3.2
96.1	Quinoline	118.0	9.5	3.4	3.7
97	di-n-Propylamine	136.9	7.5	0.7	2.0
98	Formamide	39.8	8.4	12.8	9.3
99	Dimethylformamide	77.0	8.5	6.7	5.5
99.1	N,N-Dimethylacetamide	92.5	8.2	5.6	5.0
99.2	Tetramethylurea	120.4	8.2	4.0	5.4
99.3	Hexamethylphosphor-amide[c]	175.7	9.0	4.2	5.5
SULFUR COMPOUNDS					
100	Carbon disulfide	60.0	10.0	0	0.3
101	Dimethyl sulfoxide	71.3	9.0	8.0	5.0
101.1	Ethanethiol[c]	74.3	7.7	3.2	3.5
102	Dimethyl sulfone[b]	75.0	9.3	9.5	6.0
103	Diethyl sulfide	108.2	8.3	1.5	1.0

(continued)

Table A.2. (Continued)

CODE	NAME	MOLAR VOLUME (V)	PARAMETERS HILDEBRANDS		
			δ_D	δ_P	δ_H
ACID HALIDES					
AND ANHYDRIDES					
104	Acetyl chloride	71.0	7.7	5.2	1.9
104.1	Succinic anhydride[b]	66.8	9.1	9.4	8.1
105	Acetic anhydride	94.5	7.8[a]	5.7[a]	5.0[a]
MONOHYDRIC					
120	Methanol	40.7	7.4	6.0	10.9
121	Ethanol	58.5	7.7	4.3	9.5
121.1	Ethylene cyanohydrin (hydracrylonitrile)	68.3	8.4	9.2	8.6
121.2	Allyl alcohol[c]	68.4	7.9	5.3	8.2
ALCOHOLS					
122	1-Propanol	75.2	7.8	3.3	8.5
123	2-Propanol	76.8	7.7	3.0	8.0
123.1	3-Chloro-propanol	84.2	8.6	2.8	7.2
124	Furfuryl alcohol	86.5	8.5	3.7	7.4
125	1-Butanol	91.5	7.8	2.8	7.7
126	2-Butanol	92.0	7.7	2.8	7.1
126.1	2-Methyl-l-propanol	92.8	7.4	2.8	7.8
126.2	Benzyl alcohol	103.6	9.0	3.1	6.7
127	Cyclohexanol	106.0	8.5	2.0	6.6
128	1-Pentanol	109.0	7.8	2.2	6.8
129	2-Ethyl-l-butanol	123.2	7.7	2.1	6.6
129.1	Diacetone alcohol	124.2	7.7	4.0	5.3
129.2	1,3-Dimethyl-l-butanol	127.2	7.5	1.6	6.0
130	Ethyl lactate	115.0	7.8	3.7	6.1
130.1	n-Butyl lactate	149.0	7.7	3.2	5.0
131	Ethylene glycol mono-methyl ether	79.1	7.9	4.5	8.0
132	Ethylene glycol monoethyl ether	97.8	7.9	4.5	7.0
132.1	Diethylene glycol mono-methyl ether	118.0	7.9	3.8	6.2
132.2	Diethylene glycol mono-ethyl ether	130.9	7.9	4.5	6.0
133	Ethylene glycol mono-n-butyl ether	131.6	7.8	2.5	6.0
133.1	2-Ethyl-l-hexanol	157.0	7.8	1.6	5.8
134	1-Octanol	157.7	8.3	1.6	5.8
134.1	2-Octanol	159.1	7.9	2.4	5.4
134.2	Diethylene glycol mono-n-butyl ether	170.6	7.8	3.4	5.2
135	1-Decanol	191.8	8.6	1.3	4.9
136	"Tridecyl alcohol"[d]	242.0	7.0	1.5	4.4
136.1	"Nonyl" phenoxy ethanol[d]	275.0	8.2	5.0	4.1

Table A.2. (Continued)

CODE	NAME	MOLAR VOLUME (V)	δ_D	δ_P	δ_H
ALCOHOLS					
137	Oleyl alcohol[d]	316.0	7.0	1.3	3.9
137.1	Triethylene glycol mono-				
	oleyl ether	418.5	6.5	1.5	4.1
ACIDS					
140	Formic acid	37.8	7.0	5.8	8.1
141	Acetic acid	57.1	7.1	3.9	6.6
141.1	Benzoic acid[b]	100.0	8.9	3.4	4.8
142	n-Butyric acid[c]	110.0	7.3	2.0	5.2
142.1	n-Octoic acid[c]	159.0	7.4	1.6	4.0
143	Oleic acid[d]	320.0	7.0	1.5	2.7
143.1	Stearic acid[b]	326.0	8.0	1.6	2.7
PHENOLS					
144	Phenol	87.5	8.8	2.9	7.3
144.1	1,3-Benzenediol[b]	87.5	8.8	4.1	10.3
145	m-Cresol	104.7	8.8	2.5	6.3
145.1	o-Methoxyphenol[c]	109.5	8.8	4.0	6.5
146	Methyl salicylate	129.0	7.8	3.9	6.0
147	"Nonyl"phenol[d]	231.0	8.1	2.0	4.5
WATER					
148	Water[c]	18.0	7.6[a]	7.8[a]	20.7[a]
POLYHYDRIC					
ALCOHOLS					
149	Ethylene glycol	55.8	8.3	5.4	12.7
150	Glycerol	73.3	8.5	5.9	14.3
150.1	Propylene glycol	73.6	8.2	4.6	11.4
150.2	1,3-Butanediol	89.9	8.1	4.9	10.5
151	Diethylene glycol	95.3	7.9	7.2	10.0
152	Triethylene glycol	114.0	7.8	6.1	9.1
153	Hexylene glycol	123.0	7.7	4.1	8.7
154	Dipropylene glycol[d] (mixed				
	isomers)	131.3	7.8	9.9	9.0

[a]Altered from perviously published value.
[b]Solid, treated as supercooled liquid.
[c]Values uncertain.
[d]Impure commercial product of this nominal formula.

Kirk-Othmer Encyclopedia of Chemical Technology, Stanton, A., Supplemental Volume, 2nd Ed., Copyright © 1971, John Wiley & Sons. Reprinted by permission of John Wiley & Sons.

TABLE A.3. Group Contributions to Partial Solubility Parameters

FUNCTIONAL GROUP	POLAR PARAMETER $[V\delta_P, (\text{cal ml})^{1/2}/\text{mol}]$	H-BOND PARAMETER, $v\delta_H{}^2$, cal/mol	
		ALIPHATIC	AROMATIC
—F	225 ± 25	~ 0	~ 00
—Cl	300 ± 100	100 ± 20	100 ± 20
>Cl$_2$	175 ± 25	165 ± 10	180 ± 10
—Br	300 ± 25	500 ± 100	500 ± 100
—I	325 ± 25	1000 ± 200	
—O—	200 ± 50	1150 ± 300	1250 ± 300
>CO	390 ± 15	800 ± 250	400 ± 125
—COO—	250 ± 25	1250 ± 150	800 ± 150
—CN	525 ± 50	500 ± 200	550 ± 200
—NO$_2$	500 ± 50	400 ± 50	400 ± 50
—NH$_2$	300 ± 100	1350 ± 200	2250 ± 200
>NH	100 ± 15	750 ± 200	
—OH	250 ± 30	4650 ± 400	4650 ± 500
(—OH)$_n$	$n(170 \pm 25)$	$n(4650 \pm 400)$	$n(4650 \pm 400)$
—COOH)	220 ± 10	2750 ± 250	2250 ± 250

Kirk-Othmer Encyclopedia of Chemical Technology, Stanton, A., Supplemental Vol., 2nd Ed., Copyright © 1971, John Wiley & Sons. Reprinted by permission of John Wiley & Sons.

Table A.4. Estimated Tube Counts
Triangular Pitch 3/4 (in.) Tubes on 15/16 (in.) Pitch

SHELL DIA. (IN.)	TUBE PASSES				
	1 PASS	2 PASS	4 PASS	6 PASS	8 PASS
8	43	37	26	0	0
10	71	64	50	36	0
12	107	98	81	64	0
13.25	133	123	104	85	66
15.25	180	169	147	125	102
17.25	235	222	197	171	146
19.25	296	282	253	225	196
21.25	365	349	318	286	254
23.25	441	424	389	354	319
25	513	494	457	419	382
27	603	582	542	501	460
29	699	677	633	589	546
31	803	779	732	685	638
33	913	888	838	788	738
35	1031	1005	951	898	845
37	1156	1128	1072	1015	959

Table A.5. Estimated Tube Counts
Triangular Pitch 3/4 (in.) Tubes on 1 (in.) Pitch

SHELL DIA. (IN.)	TUBE PASSES				
	1 PASS	2 PASS	4 PASS	6 PASS	8 PASS
8	37	33	0	0	0
10	63	56	44	0	0
12	94	86	71	56	0
13.25	117	108	92	75	0
15.25	158	149	129	109	90
17.25	206	195	173	150	128
19.25	260	248	223	198	173
21.25	321	307	279	251	223
23.25	388	372	342	311	280
25	451	435	402	369	336
27	530	512	476	440	404
29	614	595	557	518	480
31	705	685	644	602	561
33	803	781	737	693	649
35	906	883	836	789	743
37	1016	992	942	892	843

Table A.6. Estimated Tube Counts
Square Pitch 3/4 (in.) Tubes on 1 (in.) Pitch

SHELL DIA. (IN.)	TUBE PASSES				
	1 PASS	2 PASS	4 PASS	6 PASS	8 PASS
8	51	48	42	35	0
10	77	73	65	57	0
12	107	103	93	84	75
13.25	129	124	114	104	94
15.25	168	162	151	139	128
17.25	212	206	193	180	167
19.25	261	254	240	226	211
21.25	316	308	292	277	261
23.25	375	367	350	333	315
25	432	422	404	386	367
27	501	491	471	451	432
29	575	564	543	522	501
31	654	643	621	598	576
33	739	727	703	679	655
35	829	816	791	765	740
37	924	910	883	857	830

Table A.7.   Estimated Tube Counts
Triangular Pitch 1 (in.) Tubes on 1 1/4 (in.) Pitch

SHELL DIA. (IN.)	TUBE PASSES				
	1 PASS	2 PASS	4 PASS	6 PASS	8 PASS
8	24	21	0	0	0
10	40	36	28	0	0
12	60	55	46	36	0
13.25	75	69	59	48	0
15.25	101	95	83	70	0
17.25	132	125	111	96	82
19.25	167	159	143	127	110
21.25	205	196	179	161	143
23.25	248	238	219	199	180
25	289	278	257	236	215
27	339	328	305	282	259
29	393	381	356	332	307
31	452	438	412	385	359
33	514	500	472	443	415
35	580	565	535	505	475
37	651	635	603	571	539

Table A.8.   Estimated Tube Counts
Square Pitch 1 (in.) Tubes on 1 1/4 (in.) Pitch

SHELL DIA. (IN.)	TUBE PASSES				
	1 PASS	2 PASS	4 PASS	6 PASS	8 PASS
8	33	31	27	0	0
10	49	47	42	37	0
12	69	66	60	54	0
13.25	83	79	73	66	0
15.25	107	104	96	89	82
17.25	136	132	123	115	107
19.25	167	163	154	144	135
21.25	202	197	187	177	167
23.25	240	235	224	213	202
25	276	270	259	247	235
27	320	314	302	289	276
29	368	361	348	334	321
31	419	412	397	383	368
33	473	465	450	435	419
35	530	522	506	490	474
37	591	583	565	548	531

ESTIMATED INSTALLED INSULATION COSTS

Table A.9. Calcium Sillicate Insulation 0.016 (in.) Thick Aluminum Cover

Insulation Depth (in.)

PIPE SIZE (in.)	1	1.5	2.0	2.5
.5	9.33	11.08	12.83	14.58
.75	9.57	11.42	13.27	15.12
1	9.81	11.76	13.71	15.66
1.5	10.30	12.44	14.59	16.73
2	10.79	13.13	15.46	17.80
2.5	11.28	13.81	16.34	18.87
3	11.77	14.49	17.22	19.95
4	12.74	15.86	18.97	22.09
6	14.69	18.59	22.49	26.38
8	16.65	21.32	26.00	30.67
10	18.60	24.05	29.51	34.96
12	20.55	26.78	33.02	39.25
14	22.50	29.51	36.53	43.54
16	24.45	32.25	40.04	47.83
18	26.41	34.98	43.55	52.12
20	28.36	37.71	47.06	56.41
24	32.26	43.17	54.08	64.99

Insulation Depth (in.)

PIPE SIZE (in.)	3	3.5	4.0	4.5	5.0
1	24.78	27.58	30.39	33.24	36.10
1.5	26.11	29.02	31.97	34.95	37.97
2	27.43	30.47	33.55	36.67	39.84
3	30.09	33.36	36.70	40.10	43.59
4	32.74	36.25	39.85	43.54	47.33
6	38.05	42.04	46.15	50.41	54.82
8	43.36	47.82	52.45	57.28	62.30
10	48.66	53.60	58.76	64.14	69.79
12	53.97	59.39	65.06	71.01	77.27
14	59.28	65.17	71.36	77.88	84.76
16	64.59	70.96	77.67	84.75	92.24
18	69.89	76.74	83.97	91.62	99.73
20	75.20	82.52	90.27	98.49	107.21
24	85.82	94.09	102.88	112.23	122.18

[1]Cost basis January 1985, $/ft of pipe

Adapted from Process Plant Construction Estimating Standards, Richardson Engineering Services, Inc. San Marcos, California © 1984.

INDEX

ABSORBER, 278–282
Acentric factor, estimation of, 197
Atomic volumes for diffusion
  gases, 108
  liquids, 109

BALANCE, 65–67
Boiling point, estimation of, 197

Critical properites, estimation of, 181–197
  critical compressibility factor, 197
  critical pressure, 187
  critical temperature, 182
  critical volume, 192
CYCLONE 1, 257–260
CYCLONE 2, 261–264
Cyclone configuration factor, 250–251
Cyclone dust collectors
  correction for dust loading, 250
  nomenclature, 249
Cyclone efficiency, 252
Cyclone pressure drop, 251

Data encryption, 300
Data plotting, 1
DBLPIPE, 53–62
DCF, 165–169
Depreciation schedules, 145, 162
DIFFEQ, 23–26
Differential equations, solution to, 20
Discounted cash flow, 145, 162
Dispersion modelling, 242
DISTILL, 101–107

Emissivity constants for common materials,
  305
ENCYRPT, 302–304

EQUIL, 72–78
Equilibria, 63–95
  chemical, 63, 67
  gas solubility, 88
  vapor-liquid, 78
ESTIMATE, 140–145
EXCHANGE, 38–45
EXCOST, 173–180
Expansion factors, 283
Exponential regression, 1

Flow nozzles, 283
Free energy changes, 69
Fugacity coefficients, 70

GASSOL, 91–95
GASVISC, 221–225
Geometric regression, 1

Heat capacity, estimation of, 203–205
  gases, 203
  liquids, 204
  pressure correction for gases, 205
Heat exchangers, 32–62
  costs of, 169
  double pipe with fins, 45
  shell and tube, 32
  TEMA designation for, 171
  tube counts for shell and tube units,
Heat of vaporization
  estimation of, 197
  for organic compounds, 313
Heat transfer coefficients
  for double pipe exchangers, 46
  for shell and tube exchangers, 32, 33
Heights of transfer units in packed towers,
  273

Horizontal tanks, capacity of, 292
HTCAP, 208–215

Independent reactions, 63
INSUL, 308–311
Insulation costs, 337
Insulation thermal conductivity, 306

Jacobian matrix, 37

LIQCOND, 238–240
Linear regression, 1
Liquid density, estimation of, 225
Liquid rates in venturi scrubbers, 265
Liquid surface tension, estimation of, 235
LIQVISC, 230–235
Log-linear regression, 1

Multi-component distillation, 96

NEWTON, 28–31
Nonlinear equations, solution to systems of, 26

ORIFICE, 288–292
Orifice flow meters, 283

Packed tower hydraulics, 118
  packing factors, 119
Packed tower mass transfer preformance, 273
Parabolic regression, 1
Partial solubility parameters, 334
Pasquill stability classes, 243
PC, 190–192
Pipe insulation, economical thickness, 304
Pollution dispersion, 242
POLY, 19–20
Polynomial roots, 18
Prediction of physical properties, 181–182
Pressure drop
  in double pipe heat exchangers, 46, 47
  in packed towers, 118
  in shell and tube heat exchangers, 32, 33
  in trayed towers, 126

Process plant estimates, 136
  equipment weighting factors, 138
Project economics, 145

Reciporical regression, 1
REGRESS, 6–18
Regression methods, 1
Runge-Kutta method, 20

Sieve trays
  tray efficiency, 107
  tray hydraulics, 123
  tray nomenclature, 111, 112
SPEC, 151–162
Solubility parameters, table of, 328
Solvent association parameters, 109
Spherical tanks, capacity of, 292
STACK, 245–248
Stack height, 243

TANK, 294–300
TC, 184–187
Thermal conductivity, estimation of
  gases, 215
  liquids, 235
  pressure correction for gases, 217
TOWER, 121–123
TRAY, 129–135
Transfer units in packed towers, 275
TRYEFF, 114–118

VAPOR, 200–203
Vapor pressure, estimation, 197
Vapor pressure constants for pure liquids, 313
VC, 195–197
VENTURI, 269–273
Venturi flow meters, 283
Venturi scrubbers, 264
Venturi scrubbing systems, 266
Vertical tanks, capacity of, 292
Viscosity, estimation of
  gases, 215
  liquids, 225
  pressure correction for gases, 216
VLE, 84–88